The Journey to Dissertation Success

Are you about to embark on a research project for the first time? Unsure which data collection methods are right for your study? Don't know where to start?

By presenting the reader with a series of key research management questions, this book introduces the novice researcher to a range of research designs and data collection methods. Building an understanding of these choices and how they can impact on the dissertation itself will lead to a more robust and rigorous dissertation study.

This book is designed to direct your research choices with informative text and key questions, advice from 'virtual supervisors' and reflections from students. Lists of suggested further reading also help to support you on your journey to developing an organised and successful dissertation project.

Researchers seeking support on their journey to a successful dissertation will find this book a valuable resource.

Elizabeth Laycock is Professor of Stone Conservation at Sheffield Hallam University.

Tim Howarth is Senior Lecturer in Construction Management at Sheffield Hallam University.

Paul Watson is Professor of Building Engineering and Director of Partnership Development at Sheffield Hallam University. He was previously the Head of the Department of the Built Environment at Sheffield Hallam University.

The Journey to Dissertation Success

For construction, property, and architecture students

Elizabeth Laycock, Tim Howarth and Paul Watson

Routledge
Taylor & Francis Group

LONDON AND NEW YORK

First published 2016
by Routledge
2 Park Square, Milton Park, Abingdon, Oxon OX14 4RN

and by Routledge
711 Third Avenue, New York, NY 10017

Routledge is an imprint of the Taylor & Francis Group, an informa business

British Library Cataloguing-in-Publication Data
A catalogue record for this book is available from the British Library

Library of Congress Cataloging in Publication Data
Names: Laycock, Elizabeth, author. | Howarth, Tim, author. |
Watson, Paul (Paul A.), author.Title: The journey to dissertation
success : for construction, property, and architecture students / Elizabeth
Laycock, Tim Howarth and Paul Watson. Description: New York :
Routledge, 2016. | Includes bibliographical references and index.
Identifiers: LCCN 2015047128 | ISBN 9781138839168 (hardback :
alk. paper) | ISBN 9781138839175 (pbk. : alk. paper) |
ISBN 9781315733562 (ebook : alk. paper) Subjects: LCSH: Architecture--
Research--Methodology. | Building--Research--Methodology. | Dissertations,
Academic. Classification: LCC NA2500 .L387 2016 | DDC 808.06/6--dc23LC
record available at http://lccn.loc.gov/2015047128

ISBN: 978-1-138-83916-8 (hbk)
ISBN: 978-1-138-83917-5 (pbk)
ISBN: 978-1-315-73356-2 (ebk)

Typeset in Goudy Old Style by
Servis Filmsetting Ltd, Stockport, Cheshire

Contents

List of figures and tables

Figures

Tables

Acknowledgements

All of the authors would like to thank their friends and colleagues for their support during the writing of this book.

In addition, Liz would particularly like to thank the following people for their input: Dr Kevin Spence, Dr Rebecca Sharp, Emma Harrison, Steve Hetherington, Angela Maye-Banbury, Dave Weatherall, Norman Watts, Rionach Casey, Rob Hunt and Paul King. She would further like to acknowledge the support of her mother, her daughters and her long-suffering husband.

1 Preparing for your research journey

Introduction

This chapter presents an introduction to the book, including:

- an outline of the purpose of this book – to assist students in successfully journeying through the dissertation research process by enhancing understanding of both the dissertation process and approaches to undertaking research;
- an outline of the format of this book – one that requires students to interactively navigate the text whilst considering key theory and practice, the guidance of three supervisors and the reflections of other students upon their dissertation experience;
- an overview of the dissertation process, including a flow chart and indicative timetable;
- student reflections on the dissertation.

The purpose of this book

Researching and presenting a dissertation is a requirement of many university degree courses. It can be both very challenging and very rewarding; and for many students, it is by far the largest piece of independent research that they have encountered. The dissertation is an undertaking that demands significant commitment from the student researcher. This book has been written to support students of the built environment disciplines through their dissertation journey and to assist them in making research decisions along the way. The book outlines dissertation research processes and supports students in thinking about their approach to designing, managing and delivering their research dissertation.

A broad range of professional disciplines contribute to designing, developing, delivering and managing the built environment. These professions include: architecture, architectural technology, building surveying, civil engineering, construction management, estate management, quantity surveying and town planning. A good number of dissertation research opportunities are presented by the built environment, its history, its evolution and its many professions, practices and activities. The array of potential topics for investigation is vast and includes, but is not limited to: master planning; conceptual design; detailed design; comparative studies; materials and product innovation; legal and contractual management; planning; surveying; valuation; and the management of people, resources, costs, sustainability, design, quality, building information, environmental impact, and health and safety.

As well as a broad array of built environment research topics, there is also a variety of possible approaches to carrying out dissertation research. These include, among others: laboratory-based experiments; practice-based reviews; comparative studies; case study investigations; attitudinal and behavioural investigations; and desktop-based literature reviews, which are conducted without new primary data being collected by the researcher.

Being given the opportunity to investigate a topic that you have selected can be exciting and maybe a little daunting, with many challenges presented along the way. Without a doubt, though, doing a built environment dissertation can be a satisfying and enjoyable experience. Having the opportunity to set the focus of your research, investigate what research already exists related to your topic, determine the appropriate methods, collect and analyse data, and present it in a dissertation format can be very enriching and even the start of your own specialism within a professional field.

The format of this book

This book has been written for novice researchers, and the authors have assumed that the reader does not have extensive research experience. It considers approaches to undertaking research and outlines key concepts, definitions and requirements thereof. Furthermore, it offers an interactive means to engage with developing an understanding of some potentially daunting aspects of the dissertation challenge. Whilst the dissertation process is outlined and important aspects of research theory and practice are addressed, the book also delivers:

- a succinct introductory overview for each section;
- information in the form of text, diagrams and tables to support study of the dissertation project;
- reflections by real students upon aspects of the dissertation process;
- supervisor guidance and opinion to help the reader critically reflect upon their own dissertation;
- a series of key questions for the student researcher to consider and address;
- suggestions for further reading.

The book is structured so as to encourage the reader to move between sections and browse rather than demanding strict linear reading from front cover to back cover. It incorporates a number of features to enhance and deepen understanding of the dissertation process and the challenges associated with conducting dissertation research. The key features are designed to support you in your dissertation journey.

It is intended that this book complement the advice and guidance of a dissertation supervisor, and wider reading should always be undertaken to stimulate critical thought about performing and managing research as well as critical understanding of the dissertation process itself.

Overview of the dissertation process

Developing and delivering a built environment dissertation can be viewed as a 'project' – a project with clearly defined goals and outcomes to be delivered with only limited resources and set within a specified time frame. Recognition of previous relevant research is required, and the collecting and analysing of data needs to be carried out in valid and meaningful ways. Not only must the researcher understand the dissertation research process, they also need to be able to identify and manage research issues that arise at each stage of the process. Indeed, like any project, all aspects of the process require a great deal of thought, planning and management. Thankfully, a supervisor is allocated to help and support you through the development, planning and management of your dissertation.

Figure 1.1 outlines the main stages in the dissertation research process. Thinking about and determining a suitable topic is the very first stage in the journey; proofreading and refining the quality of the presented thesis is the final stage before submission and, hopefully, celebration.

Figure 1.2 provides an outline overview of the main considerations in managing the dissertation research process. This diagram concisely maps the key components and considerations of the dissertation journey. Importantly, it also emphasises interaction with an allocated supervisor as a valuable constituent part of the process. It should be remembered that whilst you are responsible for managing and delivering your dissertation research, your supervisor can be very helpful in guiding you and questioning your approach as you go along.

You should plan to have meetings with your supervisor throughout your dissertation research journey. Such meetings will vary in frequency and duration, but you should always try to attend with a clear purpose in mind. It is recommended that at such meetings, you be prepared to update your supervisor with regard to your progress. It can be useful to draft in advance a list of questions or issues for discussion.

In the early stages of formulating and developing a dissertation research project, you will need to meet with your supervisor on a number of occasions. It is good practice to develop a research plan and consult with your supervisor for thoughts, views and feedback regarding this proposed plan. Some universities formalise the requirement for a research plan by requiring that dissertation students be assessed on the basis of what is commonly termed a 'research proposal' or 'interim report'. Figure 1.3 provides a concise map of the process of developing a research plan and serves to emphasise that supervisor feedback is an essential and integral part of planning and developing a dissertation research proposal.

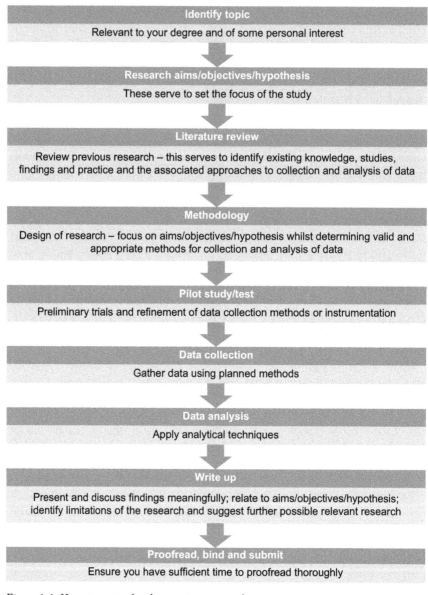

Figure 1.1 Key stages in the dissertation research process

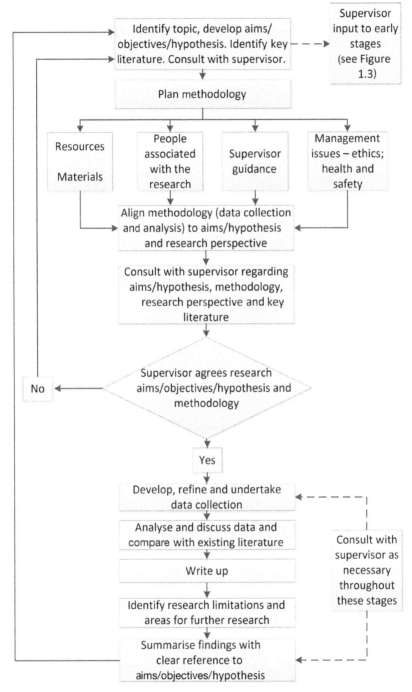

Figure 1.2 Managing the dissertation research process: an overview of key considerations

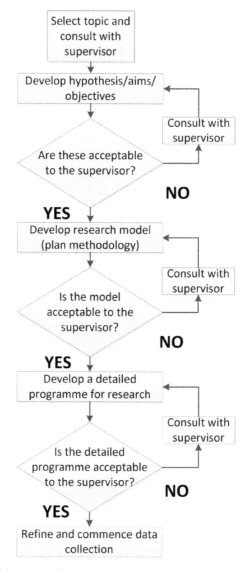

Figure 1.3 Flow diagram of the research plan development process

Dissertation research timetable

Dissertation research is commonly conducted over a nine-month period, though the timescale can vary between universities and degree courses. In some instances, the dissertation may be conducted over a 12-month period or longer. There is no one single universal timetable for the delivery of undergraduate and postgraduate dissertation research. Having said this, Table 1.1 provides an indicative timetable. This timetable assumes that the dissertation is to be conducted over a nine-month

Table 1.1 Indicative timetable for conducting dissertation research

Month	Plan of action			
1	**Decide on research topic** • read broadly • consider what interests you • current 'hot topics'			
2	Undertake broad literature review and determine research aims/objectives			
3	Refine aims/objectives, develop research methodology and ensure full ethical consideration is given to the research			
	Primary data source is people's views and opinions via interviews and questionnaires	**Primary data source is observation and fieldwork**	**Primary data source is laboratory research**	**Primary data source is workplace case study**
4	• Refine research methodology and ensure that the research meets health and safety requirements	• Refine research methodology and ensure that the research meets health and safety requirements	• Refine experimental methods and ensure that the research meets health and safety requirements	• Refine research methodology and ensure that the research meets health and safety requirements • Secure approval of employer
	Research proposal/interim report			
5	• Refine approach to primary data collection • Commence data collection • Commence literature review	• Develop approach to managing fieldwork • Commence literature review	• Experiments/ data analysis • Commence literature review	• Refine approach to data collection and commence fieldwork • Commence literature review
6	• Data collection ongoing • Literature review ongoing	• Fieldwork • Literature review ongoing	• Experiments/ data analysis ongoing • Literature review ongoing	• Data collection ongoing • Literature review ongoing
7	• Complete data collection • Data analysis • Complete literature review • Write up	• Data analysis • Complete literature review • Write up	• Complete data analysis • Complete literature review • Write up	• Data analysis • Complete literature review • Write up
8	Complete writing up and proofread			
9	**Submit final report**			

period and that there are to be two assessment submissions: an interim report/ research proposal and the final dissertation thesis.

Student reflections

A core component of this book is presentation of the thoughts, reflections and experiences of built environment students who have completed either an under-graduate or a postgraduate dissertation. Anonymous student feedback regarding the dissertation research process was collected over a four-year period. Students were asked to be frank and honest in their comments and reflections. They were also informed that the feedback would be used as *vox pop* to better inform future students about the dissertation research process. A few student reflections regarding 'the dissertation' are included below. Many such reflections are included throughout the book, and they are suitably aligned so as to be relevant to the dissertation research topics being discussed.

Student reflections regarding the dissertation

Dissertation students were asked to reflect (anonymously) upon their dissertation experiences. Here is what they said:

'Overall I enjoyed doing the dissertation even though at first I saw it as being a dreaded final year task. I cannot stress enough just how important it is to pick a topic that you enjoy as it makes it so much easier and enjoyable, even if it is hard work.'

'I have not ever written a piece of work as long and demanding as my dissertation; I'm sure that it will be the same for you. It is demanding, and for these reasons, you are bound to make mistakes or want to change certain things along the way. Keep on top of it or you will end up going with something that in the end makes little sense and just doesn't hang together.'

'I can honestly say that the dissertation was the most difficult, stressful academic challenge for me. I got lost along the way and had many a sleepless night worrying about how I was going to do certain things and wondered if I would actually get to the finish line. Now that I have, it's an absolute awesome feeling. It is well worth the stress and worry. If I were to offer advice to future students, I would say that once you know deadlines, start early, and don't put off until tomorrow what you can do today. Save yourself the time and worry because the time flies by.'

'The dissertation tested a number of my personal competencies such as my ability to manage my time. A lot of the challenge is down to how you plan and manage yourself. It has certainly developed my skills further. It also helped me to better recognise the development and progression that I have made over the course of my degree studies.'

'A dissertation is hard work but so worth it; you will feel proud when you hand it in!'

Researcher questions
Overview of the dissertation process

Now that you have read Chapter 1, 'Preparing for your research journey', you should appreciate that the dissertation research process is made up of various stages. Each stage requires careful thought, planning and management. A dissertation supervisor will support you on your journey, and a research plan will help you formulate and communicate what you intend to do and how.

1   Can you outline the key dissertation research stages?

Yes.

No. <u>Please revisit Chapter 1 and spend time studying Figure 1.1.</u>

2   Do you appreciate that you will need to develop a research plan in consultation with your supervisor?

Yes.

No. <u>Please revisit Chapter 1 and spend time studying Figure 1.3.</u>

Are you aware of the research timetable (set by your university) that you are working to?

Yes.

No. It is advisable that you consult with your university to determine the exact timetable for completion. Consider Table 1.1 and compare this to your university's requirement.

Summary

This chapter has briefly introduced the purpose and approach of the book and has provided a concise outline of the dissertation research process. Emphasis has been placed upon the need for support and guidance from a supervisor in developing and delivering a successful dissertation. This book serves to provide supplementary help and support to students journeying through the terrain of the dissertation. This is achieved by the presentation of research theory and practice, the inclusion of supervisor guidance and opinions, real student thoughts and reflections upon the dissertation research process, and the posing of key questions for consideration by you, the dissertation researcher.

2   The dissertation challenge

Introduction

This chapter presents:

- a brief introduction to the dissertation document and the student skills which are tested by the process;
- ideas to help in the selection of a topic or area for research;
- an overview of methodology and the links between this and aims and objectives;
- an introduction to the virtual supervisors who will give staff insights on each of the aspects explored throughout the book;
- student reflections on the dissertation challenge.

What is a dissertation?

A dissertation is the name given to a document written in high-quality academic language that details the student's own research. Other names such as 'research project' or 'research portfolio' may be used. While the output may not always be a bound document, this is the most common form. This document is written in a very specific style, and guidance will be given on this. Other outputs may include a short paper in the style of an academic publication, a portfolio of work, an artefact constructed after detailed research, or an oral presentation or a verbal defence of the work. Assessment may be by one or a selection of these.

Assessment varies between institutions, and there may be one or several points at which work is marked and feedback given.

The dissertation is different from much work at undergraduate level in that there is the expectation that the individual student, working largely on their own with relatively little input from academic tutors, will assemble a variety of appropriate data, synthesise and analyse this and present a conclusion. The self-directed nature of the work is often seen to be what distinguishes a graduate. The dissertation may provide evidence demonstrating a number of key graduate skills that are sought by employers (Table 2.1).

Table 2.1 Key graduate skills demonstrated by the dissertation

Skills	Evidence
Written communication	The document should be clear, concise and focused, and tailored for the audience
Commitment Motivation Enthusiasm Perseverance	The dissertation represents a large piece of work undertaken for an extended time period
Organisation	The work requires that a student can prioritise work to be efficient and manage their time well to meet deadlines
Ability to work under pressure	The process requires keeping calm and carrying on
Negotiation Interpersonal skills	Negotiation occurs between a student and their supervisor and/or research participants
Problem solving	The ability to take a logical and analytical approach to solving problems and, in particular, to be aware that there may be several approaches is necessary
Commercial awareness	Work may focus on how part of a business or the industry as a whole works or on what an organisation needs, and this may change with time
Confidence	The dissertation allows a student to demonstrate that they are confident of their ability without being arrogant or blinkered
Continuing development	Through independent work, the student can identify areas where they do not know enough, and they can then work to remedy this
Oral communication	*Viva* or verbal defence of the work demonstrates the ability to think under pressure and respond clearly and with confidence

Another challenge with the dissertation is that, unlike most other coursework briefs, it does not always follow a linear progression from a known start point to a known finish point. In many respects, a student undertaking research chooses their own start point and aims towards an anticipated outcome, which may or may not be realised. Assessment is based on a much greater collaborative effort between staff and student than most previous experiences. The student needs to communicate the outcomes they intend to achieve at the end of the process, and they will be judged against these.

The research is usually done to answer a question that frames the purpose of the research (Table 2.2). For each question that is set, there may be many different ways of obtaining the necessary data to answer the question; and one or more of these ways may be selected. These are outlined in greater detail in Chapter 6 and Chapter 7.

The student's choice of research question will depend on:

- the topic that has been selected for the study;
- what a student wishes to achieve from the study;
- practical issues relating to resources.

Table 2.2 Questions that help to frame research

Question	Research focus
What?	Describing
Why?	Explaining
How?	Exploring
When?	Identifying time(s)
Where?	Identifying location(s)
Who?	Identifying individual(s) or group(s)
Does it work?	Evaluating

Researcher question
What is a dissertation?

Do you have a clear idea of the broad outcomes of the dissertation process?

Yes. I have read the section and am ready to progress.

No. I have not read this section. I am aware that I will need to understand the process and I commit to return and read this section having journeyed further into the book.

Introduction to the dissertation process

The process of a dissertation can be considered in diagram form (Figure 2.1).

In the very broadest terms, the work will progress from initially concentrating around the topic and research question towards undertaking the research and writing up. There will be multiple forwards and backwards steps along this route, and the feasibility/practicality and relevance of the research will be considered at all stages. Further information is given in Chapter 3.

Initial requirements of dissertation research

When carrying out research, a student is advised to consider, where possible, selecting a topic or area in which:

- there is a personal interest;
- it is feasible to collect data;
- the personal skills of the researcher align with the methods of data collection being considered;
- supervising academic staff have relevant expertise.

Within each of the points in the list above, it is possible to further narrow the possible choices and refine the area of focus (Table 2.3).

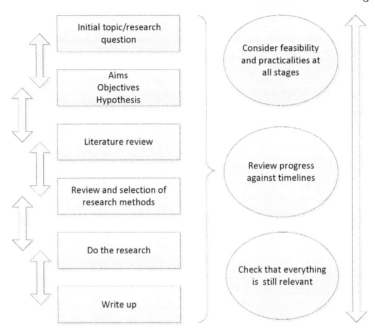

Figure 2.1 The flow of the dissertation process

Table 2.3 Focusing the topic area

Aspects to consider	Suggestions
Topic of personal interest/current relevance	An issue arising from work or placement experience
	An issue which aligns with student career aspirations
	A topic of general interest that is currently discussed in literature
	An area where a new initiative or new legislation has been introduced
	A topic relating to differences in attitudes/practice.
	An area where there seems to be disagreement relating to current practice or innovation
Feasibility of collecting data	The study can be narrowed down geographically or by careful sample selection
	The source of data should not lie beyond the researcher in terms of budget or accessibility
	The intended research has no major ethical issues
Personal skills of the researcher align with the methods of data collection considered	The student should feel able to carry out the required methods, and these should lie broadly within their skill set and preferences.
	The data analysis which will be required should be feasible for the student
Academic staff have expertise in the area	Staff expertise (in both the topic area and the research methods) can offer significant advantages in the early stages of a project. However, given the personal nature of some dissertations, it is possible that the supervisor will not be an expert in the field and will only be providing expertise on the processes of research.

Table 2.4 Problems with high or low volumes of literature

Availability of literature	Problems
Large volumes of literature	• Reading and summarising will take a lot of time, much of which may be spent reading similar documents • Danger that the topic has been too well covered to provide any insights from dissertation research (where primary data are to be collected) • Referencing becomes a logistical issue
Small volumes of literature	• Lack of underlying secondary data on which to build • Consider if there is a reason that this topic is not widely discussed • Consider if the literature is there but not readily available

Further information can be found in Chapter 3.

Some institutions provide a list, based on staff research interests, of potential titles or areas for undergraduate dissertations. If this is the case, it is advised that a student should determine whether these are fixed or whether the titles are indicative, having flexibility to be changed slightly.

The dissertation involves extensive self-directed study. As such, it is strongly advised that a topic of limited scope or one which is likely to become 'boring' is avoided. Initial assessment of the literature should be undertaken to clarify the amount of information available for the literature review. Problems are posed by having either a very large volume of available literature or a very small amount (Table 2.4).

Researcher question
Initial requirements of dissertation research

Do you have a clear idea of the ways in which you might start to consider and narrow your topic?

Yes. I have read the section and am ready to progress.

No. I have questions which have not been answered and need further information before I can progress. I recognise that this is a normal part of the initial phases and will return to this section having undertaken further reading.

Methodology: data collection and analysis

Dissertations contain a section on methodology, which is an explanation and justification of the ways in which information will be collected. The methodology needs to make clear to the reader how the objectives will be met and how these will allow the aims to be achieved or the hypothesis to be answered. The methodology commonly addresses:

- the range of available methods;
- the method to be used;
- the strengths/weaknesses of the chosen method;
- the contingency plan (this may be an alternative research method).

The methodology has to satisfy criteria of validity and reliability. The choice of methodology will depend on:

- the topic chosen;
- what outcomes are anticipated;
- consideration of practical issues relating to resources;
- the contingencies identified.

A contingency offers a way of continuing the work if the first attempt at data collection does not succeed. While it is possible that this will not be needed, it is recommended that consideration be given to contingencies at this early stage as part of the assessment of feasibility.

It is often helpful to consider how the whole project will flow and which aims and objectives will be met by each method used. This may be an iterative process, but ideally within a few weeks of the start, a student should have enough of an understanding to communicate the proposed title; the initial aims and objectives; and a list of methods which will be linked to each objective which will allow completion.

The methodology can then be presented as a flow chart showing the linkages between aims and objectives and the ways in which these will be achieved (Figure 2.2). The literature review should also be placed into this plan. Having prepared the chart, planning requirements and timescales can be considered.

In summary, the methodology provides the following information:

- a justification of why the method(s) selected are appropriate to the research question;
- the possible weaknesses of the method(s);
- alternative method(s).

It is unlikely that one single research methodology is perfect for any research question. There are always advantages and disadvantages to be balanced and considered. The level of data which could be gained, for example, by undertaking a detailed questionnaire and a wide selection of participant interviews might be the best method, but this may not be achievable within the timescale for the dissertation. Acknowledging any disadvantages your methodology has and the methods used to minimise these may be useful.

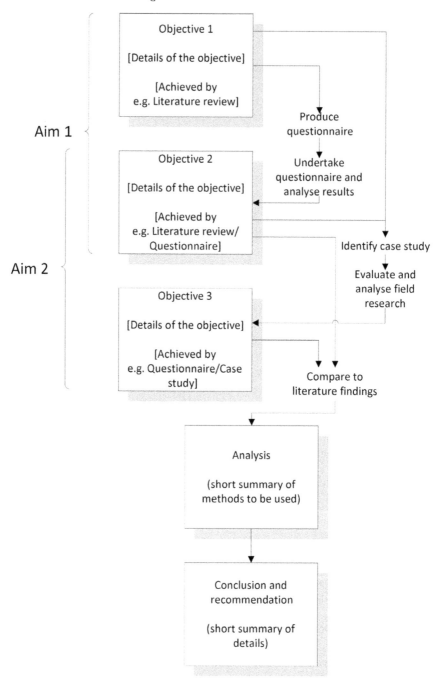

Figure 2.2 Link between aims, objectives and methods

Researcher question
Methodology: data collection and analysis

Do you understand how a methodology demonstrates how the project will proceed?

Yes. I have read the section and am ready to progress.

No. I have questions which have not been answered and need further information before I can progress. I recognise that this is a normal part of the initial phases and will return to this section having undertaken further reading.

Introduction to the role of the supervisor

An academic supervisor is a member of staff with knowledge of the process and the ability to guide a student. In some cases, a student is able to select a supervisor. Here, the academic's subject experience and research expertise and experience should be considered. Students are also advised to consider personal aspects when making their choices. The best research often comes from situations where students and supervisors get on well as people, communicate openly and honesty, and have clear, shared goals. Where this personal connection does not exist, the relationship can still succeed as long as the expectations of both staff and student are transparent and communicated effectively. The research undertaken has to be realistic in scope. A supervisor will have a clear idea of this from their previous experience. A supervisor will also be able to give advice on the timescales for research; the level of feedback available; and any further sources of support.

The student should take ownership of the supervisory process. Regular meetings will allow greater control over the project through reporting and reflecting on progress at regular stages.

Introduction to the virtual supervisors

When students compare experiences of their supervisors with other students in their peer group, they are often surprised to find that these appear to be quite different. This should perhaps be expected as a number of factors are at play: the nature of the topic, the nature of the student, the nature of the supervisor, and the interactions and expectations of all involved. To illustrate the different views that exist, the voices of three 'virtual' supervisors are used; these provide an insight into the range of advice that might be expected from different supervisors.

Introduction to the virtual supervisors

Supervisor A

I have some 20 years of experience of supervising undergraduate and post-graduate students to successful completion of their dissertation research. I have supervised hundreds of built environment–related dissertation students, and I have usually found supervision to be a very enjoyable experience. It is certainly pleasing and rewarding to see students express real interest in developing and undertaking small-scale research projects.

Supervisors, like students, vary in their approach to managing and delivering dissertation supervision. Historically, supervision was an activity that predominantly involved the student and supervisor meeting face-to-face. This is not always the case now. Technology now enables a 'mixed-mode' approach to dissertation supervision. Whilst some supervisors prefer to maintain supervision as purely a series of face-to-face meeting encounters, some supervisors, like me, may prefer to supervise by a *mixture* of: face-to-face meetings; small-scale group supervision discussions; email; telephone conversations; and video conferencing. I find the need for this mixed-mode approach is heightened when supervising students who are studying at a distance or part-time. Long gone are the days when supervision was limited to being an on-campus activity. For students who study at a distance, by part-time mode, or who attend for block-delivered lectures, access to supervision is no longer a major concern. Modern technology has greatly enhanced access to the supervision process.

For me, it is important to establish at the start of the supervision process the means by which supervision will be provided. This helps to ensure that expectations of both the student and the supervisor are clear and well aligned with regard to how supervision is to be conducted.

Regardless of a supervisor's preferred mode of supervision, in my experience, all supervisors share a common aspiration: the desire to see their students succeed. I certainly find it highly satisfying to see students developing enhanced understanding and confidence regarding research. Guiding students to deliver interesting and well-organised dissertations can be a very pleasing activity for supervisors.

Supervisor B

I've been lecturing for nearly 20 years and dealing with student dissertations and research at various levels throughout that time. For the first few weeks, I encourage students to try and decide on a topic or topic area. This can be one of the most challenging points as students often struggle to commit themselves to just one idea. The next phase is to almost simultaneously start considering where the topic might be focused, how the data might be collected and the effects this will have on the analysis techniques to be used. Students come from different backgrounds and have diverse preferences, so after just a week or so, their progress is often markedly different. My advice to students is that if you find yourself slightly adrift then talk to someone. Staff and colleagues are obvious sources of support. Once you get a supervisor, they should be your main point of contact and will work with you.

It's really important for me to see students face-to-face as I prefer to use sketches and diagrams to explain parts of the process. When I'm meeting students in person, I prefer to throw ideas between us and then try and find a path through the topic that will allow the student to get a clear understanding of the topic and their approaches and perhaps start to think about some of the pitfalls. I usually encourage them to plan for the worst and hope for the best to happen.

I do realise that not all supervisors are the same and that sometimes we supervisors might give different advice. In my view, that is because staff, students and projects are all individual, and the combination of those factors alone can lead to dissertations with a seemingly identical title or area producing quite different outcomes. These variations are not negative, but there is no doubt that they can be disconcerting where two students suddenly find themselves getting apparently conflicting advice from supervisors.

The dissertation journey is very personal, it is hard work, it can be a little bit (or very) frightening at times, and it can be quite lonely. I think knowing that someone else, particularly your supervisor, understands that can be really helpful. At the end of the day, I hope my students enjoy their finished project as much as I enjoy watching them accomplish it.

Supervisor C

I've been lecturing for well over 20 years. I have an extensive experience of my subject and of research and I bring this expertise to my students. Projects that are successful are generally the result of careful consideration and reflection, and the collection of good-quality data and appropriate and detailed analysis are paramount. There has to be clear engagement from the student from day one; and as long as suitable frameworks, timescales, targets and contingencies are in place, the project will almost always be successful. A good dissertation is a source of considerable satisfaction to both student and supervisor. We understand the process because most, if not all, supervisors have done at least one, and often several, such pieces of work at various academic levels. I urge any incoming student to use the staff to their fullest by listening carefully to their advice, which is born from experience. Supervisors want you to do well but cannot guarantee you that success unless you are willing to commit to the project and do the work.

Student reflections on the dissertation challenge

'The dissertation was the most interesting, worthwhile and satisfying module on my degree. Also the hardest.'

'You get out of a dissertation what you put in. It's hard, unrelenting work, particularly if you are working while you are studying. To do it well, you have to fully embrace your topic – you have to live and breathe it.'

'Meet the targets set and take heed of the milestones. In other parts of my degree, I put things back until the last minute. With the dissertation, I felt that I couldn't do that; and more importantly, I didn't want to.'

'I put the effort in and I was really proud of my work. Seeing the final printed document showed me just how much I put into it.'

'Doing the dissertation was extremely helpful. Not simply giving me answers but actually making me think about my subject, which helped me submerge myself into the topic.'

'My tutor told me there would be times when you might feel like you are drowning, and she wasn't kidding! The dissertation process is hard but one of the most rewarding things I have ever done.'

'I think it's fair to say that the Dissertation module epitomises what an 'academic' degree should consist of; it focuses on a balance of reasoned argument, establishing existing themes, proving such points with valid data and allowing you to justifiably question the status quo. The nature of my degree is vocational and therefore there should be a certain level of demonstrating technical competence, but academia extends to more than this and thankfully this dissertation finally recognised that!'

'It was a very daunting experience, especially at the start when I didn't fully understand the expectations of a dissertation. But as time passed, everything became clearer, and I actually enjoyed doing the work.'

'Timekeeping was my downfall. Putting off your dissertation and saying "I've got plenty of time" was a mistake as I found I had lots to do in little time, especially when I had other pieces of coursework due in.'

'It would appear after the event that a dissertation is NOT four 3,000-word essays bolted together as I thought beforehand! It really is quite difficult to manage your time and produce the quality needed.'

'The best part of doing a dissertation wasn't doing the work or even handing it in. It was picking up the document after a few months and looking at it and thinking "Did I really do work as good as this?" I didn't think I was capable of producing something that looked so professional and I was really proud of my achievement.'

Summary

This chapter has considered the key outcomes of the dissertation process and the requirements for successful completion. The need to focus the research question and to reflect on the feasibility and practicality of the intended data collection throughout the research is highlighted. The linkages between aims/objectives/hypothesis and methodology were discussed in order to ensure alignment between these and the overarching research question.

3   Doing the dissertation

Beginning the journey

Introduction

This chapter introduces the initial terrain of the dissertation journey and presents:

- key considerations in the commencement of a dissertation research project, including selecting a topic, formulating a title, and developing a hypothesis and null hypothesis, aims and objectives;
- the early-stage activity of writing a research proposal;
- guidance from virtual supervisors on choosing a topic;
- student reflections regarding selecting a topic and formulating a dissertation title;
- suggested further reading.

Setting out on the dissertation journey can sometimes be daunting and somewhat overwhelming for the novice researcher. It can be reassuring, though, to know that you are not the first person to set out on this journey and that you have the support and guidance of an experienced academic supervisor. Furthermore, you have the help and support of this text, your own guidebook. It will help you make sense of the terrain ahead and the considerations and decisions that have to be made along the way.

When setting out on any journey, it is always wise and sensible to plan the route ahead. This chapter serves to help you plan and begin your dissertation research journey. As such, this chapter provides an outline of key dissertation stages and offers a useful route planner for the commencement of your dissertation journey. In assisting you to begin your dissertation, this chapter seeks to help you better understand and manage the initial challenges ahead by drawing attention to some key considerations of the early journey.

Stages in the research process

It is useful and worthwhile to view the dissertation process as a 'project' that has clear objectives and a discernible start point and end point. There are five key stages in the research process, these being:

Researcher question
Beginning the journey

You should be prepared to commit sufficient time and effort to your dissertation. The dissertation journey can be a challenging one. Have you read and considered 'Preparing for your research journey' (Chapter 1) and 'The dissertation challenge' (Chapter 2)?

Yes. I have read these chapters and am ready to begin the journey.

No. I have not read both of these chapters.

It is advisable that you do read them before commencing your dissertation.

---

1. defining the problem or topic of investigation
2. considering and determining the methods of investigation
3. collecting information (data)
4. analysing the data
5. formulating conclusions and writing up.

The project journey involves the systematic undertaking of each of these stages. The careful consideration and timely achievement of each stage is critical to the success of any dissertation project.

Some student researchers may choose to see this process as a sequential run of events whereby each stage is addressed in turn. This can be the case but is not always so. Often there is overlap and interplay between each of the stages.

Supervisor guidance: stages in the research process

Guidance from Supervisor A

Systematic investigation needs to be the student's focus when undertaking the dissertation. My experience tells me that a structured approach is necessary. At the outset of my supervision, I ask my research students to be aware of the fact that there are various stages to a research project. The first supervision meeting with a new research student involves a discussion regarding the five key stages of dissertation research. In my experience and opinion, the first stage – that of defining the problem or topic of investigation and determining the research aims and objectives or hypothesis – is always the most difficult for students; though this may not always appear to them to be the case. Once the topic has been established, it is possible to begin investigating, considering and aligning the approaches to data collection and analysis, and to begin preparing the initial research proposal.

Guidance from Supervisor B

I supervise students who do dissertation work both inside and outside the laboratory. For the laboratory work, it's relatively easy to help students refine their initial ideas into something which will be viable and allow them to collect enough data to deliver a research project. On the other hand, I often get students who need a bit of help and guidance. I usually recommend that they start with the literature – find out what has been done and see what ideas they can collect about what they might want to test.

The main issues are students either trying to be much too ambitious or, conversely, being much too limited. I'd say that students need to get an early idea about what might be involved and what resources they might need. Sometimes students need to liaise directly with technical staff and perhaps even have small pieces of equipment built up for them. Obviously there are time issues with this, particularly since there could be several other students doing research, all crying out for technical time. It's important to realise this and start early so that you have every opportunity. Even with laboratory work, it's helpful to consider a contingency plan – what you might do if things don't work out as you had planned for. You might also need to remember that materials or equipment might have to be bought and that sometimes this might mean purchase orders having to be raised and signed off, causing more delays. You should just be aware that even with a practical project, there are elements of negotiation and resource control.

Guidance from Supervisor C

Research is a highly structured process that lends itself to a project management approach. The Deming cycle of 'plan, do, check, act' is highly appropriate to all stages of a dissertation research project. It is essential that research students read and plan before consulting their supervisor and refining and acting upon their plans. I expect students to understand that a research project is made up of a number of interrelated stages from deciding a topic to determining aims and objectives or hypothesis through selecting and aligning research methods with the investigation, carrying out data collection, analysing and discussing results, and presenting findings.

Only when students are aware of the key stages in planning and doing research can they begin their dissertation journey in a confident and informed manner.

Researcher question
Stages in the research process

Do you have a clear understanding of the key stages of the research process?

Yes. I have a clear understanding and am confident that I am ready to begin thinking about selecting a topic.

No. I am not confident that I have a clear understanding of the key stages of the research process. I think I may need to develop or strengthen my appreciation. It is probably useful that I revisit 'Stages in the research process' before commencing my dissertation journey.

Selecting a topic

There is no secret formula to determining and selecting a topic for your dissertation. There is, however, one underlying constraint: the topic must be relevant to the professional context of your degree.

Identifying and defining a problem or topic for investigation is the departure point for all dissertation studies. Only when a topic or problem has been initially established can the research process move forward.

The built environment offers a great wealth of research topics. As noted in Chapter 1, the scope of research topics is broad and encompasses a range of activities and practices including: master planning; conceptual design; detailed design; comparative studies; materials and product innovation; legal and contractual management; planning; surveying; valuation; and the management of people, resources, costs, sustainability, design, quality, building information, environmental impact, and health and safety.

Built environment research topics are certainly not limited to the above list though. This simply provides an indication of some possible broad areas within which research might be carried out. If you do have difficulty in deciding or selecting a topic of study, a set of lists may help you in this process.

- topics/subjects you have studied during your degree course that you found particularly interesting;
- topics/subjects that have recently had a high profile within the media;
- topics/subjects that are relevant to your workplace (if you are currently employed in a relevant job or have been out on placement);
- topics/subjects that you have read about and find interesting but have not studied during your degree course;
- topics/subjects closely related to the requirements of a relevant professional institution;
- topics/subjects suggested by a tutor;
- topics/subjects that you think you will enjoy researching.

The very fact that you are making lists can help you to refine your thoughts regarding your topic choice. Revisiting the lists can prove a worthwhile endeavour as you filter and begin to exclude topics. To some, the choosing of a topic is very easy, whilst to many, the process of selecting a research topic is a challenging undertaking. It is something that needs to be done before you can start your research journey.

Student reflections on selecting a topic

Students were asked to reflect (anonymously) upon their dissertation experiences. They were told that their responses would be shared with other students in order to better inform and assist them in their dissertation journey. Listed below are the comments and advice that the research students provided regarding selection of a dissertation topic.

'The best advice to dissertation students just about to start is that you have to fully embrace your topic if you are going to make a success out of it. You've got to live and breathe the topic.'

'Make sure that you select a subject that interests you in order to keep you stimulated.'

'Make sure that you choose a topic that you will be interested in; give it a lot of thought! It is a lot of work over a relatively long period of time, so if you don't enjoy your topic, it will be very difficult to do a good job.'

'If possible, select a subject area that you may benefit from in future roles and that will improve your job performance.'

'Even if you have genuine interest in a topic, make sure there are plenty of resources available.'

'Think hard and do prior research about which subjects interest you because a few months in you may realise you have picked one that doesn't provide enough information to write a full dissertation on.'

'Think about taking the opportunity to do something new. Something not covered in your course. I did a topic that wasn't really covered in the course syllabus but was relevant to my degree. This probably helped me enjoy the dissertation study even more.'

'Choose a topic that interests you and has information that is readily available.'

'Look through up-to-date magazines and journals and consider topics that are recent and current. This could be done online. Looking at current issues may help you discover what interests you and may help you to get a general topic idea.'

'In my experience, when selecting a topic you should really think about:

- your own areas of interest – this can help to prolong the motivation you have to help see the dissertation through;

- the potential contacts, resources and business organisations that may be available to you for data collection;
- the supervisors you could potentially be allocated with regards to your topic.'

'Do plenty of research and reading when choosing a topic. I found it useful to look at previously completed dissertations in the library.'

'It's good if you can begin to look for your topic at least one month before you start the dissertation module. Pick something that you find interesting. I did mine based around the study of management practices on a large commercial project.'

'Choose a subject area. Define the sort of issues that you would like to explore in a few words. Then try to fill in the gaps and join the words together.'

'Make sure your topic is something you're already into or something that you are itching to know more about. The choice of a wrong topic or something you have little interest in makes no sense. The dissertation becomes a major part of your life, and if it doesn't interest you, it will be like trying to eat Brussels sprouts at Christmas (if you hate sprouts like me).'

'My advice is that you should be individual and not follow the trend. Pick a unique topic to maintain interest and stop you from losing motivation. Choose something you are interested in and that there is an abundance of information on, but don't take too long about your choice.'

'It is important to ensure you focus the topic in an area where you are confident of getting the data, such as industry contacts. This will make it more likely that you can collect primary data. Don't pick a topic where getting hold of the resources will be impossible. You are just making it more difficult.'

'Consider, when choosing a topic, where the data will come from and how you will collect and analyse it. I set up all my spreadsheets before I sent out any questionnaires, which took a lot of stress out of the analysis of the data.'

'I would suggest a student spend time reading through past dissertations, just so they are aware of what they need to do. Check though if they are the best ones or just a random set so you know what you are aiming for.'

'Don't be afraid to admit that you have chosen the wrong topic early and change. You will need to dump some of your work, but that is much better than having a project that you know will go badly to live with for the next year.'

Supervisor guidance on selecting a topic

Guidance from Supervisor A

When considering or finding and defining a topic or problem for inquiry, I always suggest that it should be of some interest to the student researcher. Sometimes students ask, 'Can you give me a topic?' and I always respond: 'What interests you? Why? Are there any current issues in the industry or in your discipline that are in the news at the moment? What do you see yourself doing in three, five or ten years' time? Maybe think about these things and go away and set aside some time to search and read about potential research topics and/or problems.'

Once a topic or problem has been tentatively identified and briefly discussed, I ask my students to undertake further initial background reading, sometimes called a 'preliminary literature review'. I ask my students to read and read and to consider what has already been investigated, discovered or written about the topic? Who have they tentatively identified as being the key persons already associated with researching the topic or problem? Is there an existing published research investigation that they wish to mirror or repeat on a suitable scale?

In my experience, doing all of this helps and enables the formulation of the research aims and objectives or hypothesis. Only when this is done, can the student begin to consider developing the research methodology. Having said that, aims and objectives are commonly revisited and revised or refined when it is realised, usually during supervisory meetings, that the proposed aims and objectives are somewhat overly ambitious when time and cost constraints are considered.

Guidance from Supervisor B

I would always advise students to play to their strengths. That often causes issues as students are not sure what their strengths actually are. If in doubt, make a list! Write down the topics you have enjoyed to date as part of your degree, things that you might have encountered as part of work experience or shadowing, or related topics in the news. At this stage anything goes.

Once you have your list, systematically go through it and ask yourself questions:

- What has changed with this area recently (for example, legislative changes)?
- Are things done differently (for example, between companies)?
- Is there believed to be a problem with how this works currently/does

everyone agree with how this is perceived (for example, within or between a group of people reacting to an item)?

- Is there something in here I don't know the answer to (for example, how a material behaves and why)?
- How could this be determined? (This question links to research methods. You need to get a handle on these and consider at an early stage whether the project is viable.)

Or there may be other questions that will help you. Although it is time-intensive, going through these will start to focus your attention down to a few areas.

At this point it can be helpful to bounce ideas around with colleagues or friends, but remember to take advice with a pinch of salt. Sometimes well-meaning people will try and 'give' you a topic, and this is not always a great idea. It can be very difficult to politely refuse a topic; or for a student struggling to find a focus, it can be easy to grab on to this and keep hold, even beyond the point where the topic is not viable and is dragging them down. By all means talk to people and get ideas, but be careful to make sure that the topic is something you can live with for the next nine months or so.

What you are going to do really has to be based on this decision, so it's one of the most important things to get right. Spending a few days really thinking about your topic can save an awful lot of heartache later on.

Once you have the topic, you need to work out what you are trying to do overall. Write this down. You can worry about formalising your aims and objectives later, and these may change.

The viability of a project is a key issue; make sure that you consider this before you commit to a topic.

Guidance from Supervisor C

The topic you select should be something that you are interested in and something that you want to learn more about. It also needs to be relevant to your professional discipline. You should not choose to do something because you think it will be 'easy'; conversely, you should not choose a topic because you think nobody has ever studied the topic before and it will bring new knowledge, understanding and insight to the world. Such research would constitute PhD-level investigation with scope and resource demands beyond that of an undergraduate or postgraduate dissertation project.

Try making a shortlist of possible topics, and then try to identify just what it is that you might look to research with regard to each topic. Are you being too ambitious? Is what you are proposing to do a viable research project for the timescale and resources that you have?

If no topics 'jump out at you' then try looking at previous dissertations and journal papers for ideas and inspiration. Maybe run your ideas past a tutor or, as a last resort, ask a tutor if they have any topics, questions or small projects that you could research.

Do not, though, choose a topic based upon who you think might be your supervisor for the chosen topic. This is entirely the wrong way to go about setting off on your dissertation journey. Your interest in the topic and its potential relevance to your future professional practice should always be the prime factors in choosing your topic.

Researcher question
Selecting a topic

Do you have a topic in mind? One which you have initially investigated and gathered some preliminary information about?

Yes. I do have a topic in mind.

No. I do not have a topic yet

The dissertation title

The title of the dissertation should give indication as to the nature or content of the research. It should not be too long and should not contain any grammatical or spelling errors.

The title printed on the front cover of the dissertation may differ from the title you chose to allocate to the research early on in the study. Refining the title can be good practice so that the title printed on the thesis submission concisely reflects the completed project. When formulating a dissertation title, the researcher should consider some key points, these being:

- Does the title clearly align with the research topic?
- Does the title contain words that are not necessary or are misleading?
- Does the title on the front of the thesis accurately reflect the contents of the document?

Words and phrases commonly found in dissertation titles include:

- investigation of;
- a comparison of;
- a study of;
- analysis of;
- an assessment of;
- case study evaluation;
- critical evaluation;
- an appraisal of;
- modelling the;
- a determination of;

One method for developing a draft title is to write down:

- a few keywords that describe the topic/area of study;
- what you are measuring/what the focus of your work is;
- what the outcome of your work will be (what are you comparing/describing?).

EXAMPLE 1

1. Open plan/traditional office environment
2. Employee satisfaction
3. Comparison of attitudes

Possible first draft title:
   A comparison of employee attitudes toward open plan and traditional office environments

EXAMPLE 2

1. Lime mortar additives
2. Frost durability
3. Evaluation

Possible first draft title:
   Evaluation of hemp fibre additive to lime mortar for enhanced frost durability

---

Student reflections on the dissertation title

Here is some advice regarding the formulation of your dissertation title. The advice is provided by students who have recently completed their dissertation research.

'I thought of four possible titles early on in my research. My advice to others is that once you've chosen a broad topic, try to get a range of slightly different titles and discuss them with a supervisor or another lecturer.'

'I asked lots of people for advice on choosing a title but then got confused because they were in such different topic areas. Working out which title to use was quite hard. I now think it is probably better not to ask other people about titles until you have a pretty clear topic area.'

'The first thing I did was to decide the title of my research. This really helped me define my topic, though I changed my title a little as the dissertation progressed.'

'I left writing my title until the last thing. I did have a provisional title throughout my research. The final title was a little more refined and polished.'

'My title was suggested by my supervisor. I struggled to define one topic for study and my supervisor helped me to think about combining two topics into one dissertation by suggesting a possible title.'

'If you are using aims and objectives, it is sometimes difficult to write a title that isn't the same as one of them. My title ended up being very similar to my research aim.'

'I think there is a lot of emphasis on writing a title, and I spent too long on this aspect early on. I think that spending more time doing background reading and clarifying my topic would have better helped me formulate my title.'

Supervisor guidance on the dissertation title

Guidance from Supervisor A

My advice to students is always 'don't worry too much about the title of your dissertation until you are writing up'. The title will inevitably change a little as the research investigation progresses. It is necessary to have a preliminary title, sometimes referred to as a 'working title' or 'provisional title'. Ensure that you do not get too constrained or spend too much time considering and reconstructing your chosen working/provisional title.

The working/provisional title should embody the main thrust of your dissertation thesis and should serve to draw the potential reader in. It should be concise and should relate to the topic of your research.

Getting the title right is clearly very important when you are ready to print and submit your work. There is no one secret formula or method for doing this. I think it can be very worthwhile, when considering the title, to prepare a few possible alternative titles and discuss them with friends and colleagues. Make sure the title conveys in some way what your dissertation research is about. Be careful and precise in your choice of words; these are the words that may sit prominently on your bookshelf for years to come.

Guidance from Supervisor B

Students can get too hung up about their title and worry about that rather than getting on with actually doing the research and refining the topic area. One exercise that sometimes helps is to come up with five keywords that summarise what you are trying to do and then play around with the order of these until you find one that you are comfortable with. If you still cannot come up with a title then at least try and communicate the area that you are working in; use more words than you need to if necessary. You can refine your title as you progress through the work. You can use this as an initial provisional title.

You should make sure that the title on your work at the end matches the contents of the dissertation you hand in. Use words that communicate expectations clearly and try and keep it reasonably brief but not so short that the reader doesn't have any idea what is in the book.

Guidance from Supervisor C

Do not get too hung up trying to craft a perfect title right at the start of your dissertation research. The title is important but does not need to be concluded until the final stages of the dissertation process. Too much time can be wasted early on in the dissertation process on trying to capture an elegant summary of your research. Write down some keywords and put together a 'working title' to be used whilst the dissertation research is developing. It is important that your title reflects your research work; it just doesn't need to be decided and finalised until late in the dissertation process.

When you do decide your title, do not make it extremely long and wordy. Keep it concise and relevant.

Researcher question
The dissertation title

Do you have a provisional title for your research project?

Yes. I have a clear outline understanding and am confident that I am ready to begin selecting a provisional title.

No. I do not have a provisional title yet. I recognise that I need to give a little more thought to this and I need to establish a provisional title for my research work.

The hypothesis and null hypothesis

A hypothesis is a stated proposition which is tested or evaluated using data collected within a research investigation.

EXAMPLE 1

Adding material x to a concrete mix will enhance the compressive strength of the concrete.

The result of analysing the research data will be to either prove or disprove the hypothesis. The use of a hypothesis can provide research with a specific focus. A null hypothesis is the opposite of the stated hypothesis. The null hypothesis of the hypothesis proposed in Example 1 is stated in Example 2.

EXAMPLE 2

Adding material x to a concrete mix will not enhance the compressive strength of the concrete.

Therefore, when a research investigation proposes both a hypothesis and a null hypothesis, the research outcome will be to prove either the hypothesis or the null hypothesis.

Use of a hypothesis or null hypothesis is not a requirement of all dissertation research; rather, it is an approach that is common within research based on experiments.

Research aims and objectives

A research aim is the stated core focus of a research investigation. It is a concise summary of what the research intends to investigate or achieve. As a statement of intent, the aim sets the scope of the research and can be written in broad terms.

A number of verbs are commonly used when expressing the aim of a research project; these can include, but are not limited to:

- assess
- appraise
- define
- describe
- determine
- develop
- establish
- evaluate

- examine
- explain
- explore
- identify
- investigate
- monitor
- test.

It is very useful to state the aim of your research when friends, family, colleagues and tutors ask the question 'what is your research about?' Stating the aim of your research provides a relevant and very succinct answer to this question.

Examples of dissertation aims

The aim of this dissertation is to monitor and compare energy usage on two construction sites.

The aim of this dissertation is to investigate and describe the design processes of a small architectural practice.

The aim of this dissertation is to determine the benefits of and barriers to the use of BIM by tier 2 contractors.

The aim of this dissertation is to undertake a case study examination of the use and success of apprenticeship schemes within a construction organisation.

The aim of this dissertation is to examine the effects of adding pulverised fuel ash to various concrete mixes.

Research objectives

Research objectives support the research aim and provide specific detail as to how the research aim is to be achieved.

There is no specific 'best' or 'appropriate' number of research objectives for a dissertation. It is most important that the objectives are clear; that they relate to and support the aim; and that they are realistic and achievable. This cannot be overemphasised as the objectives serve as a 'signpost' for your research journey. They should succinctly express how the research aim is to be met. It is all too common for dissertation students to be overly ambitious in their dissertation aspirations, drafting a list of objectives that is much too long and which is unachievable.

Examples of dissertation objectives

EXAMPLE 1

The aim of this dissertation is to explore the current and future use of apps on construction sites. The objectives of this research project support the research aim and are as follows:

- to undertake a case study investigation to determine the apps currently used by a UK national contractor;
- to review literature to determine the current use of apps on construction sites;
- to identify the key benefits and barriers associated with using apps on construction sites;
- to determine the views of UK construction contractors and IT consultants regarding the potential future use of apps on UK construction sites.

EXAMPLE 2

The aim of this dissertation is to examine the effects of adding pulverised fuel ash to various concrete mixes. The objectives of this research project support the research aim and are as follows:

- to calculate the effect upon compressive strength of concrete when pulverised fuel ash is added to concrete mixes;
- to explain the effect upon the concrete slump test when pulverised fuel ash is added to concrete mixes;
- to determine the effect upon the tensile strength of concrete when pulverised fuel ash is added to concrete mixes.

Writing a research proposal

A research proposal is a document that presents preliminary planning for a research investigation. Developing a research proposal is a sound and sensible early-stage activity when undertaking a dissertation research project.

Writing a research proposal requires:

- a research topic or problem to have been chosen and defined;
- the determination of an aim and research objectives or a hypothesis;
- an identification of key literature and previous research regarding the topic of investigation;
- an outline of how you propose to collect data – identification of which data collection methods are to be used;
- an outline of how you propose to analyse data – identification of which methods of data analysis are to be used.

It is common for research proposals to consider and address potential health and safety issues (see 'Health and safety' in Chapter 7) and possible ethical issues (see 'Ethics' in Chapter 7). Universities invariably have robust processes for considering the health and safety and ethical issues associated with research projects, including those conducted as undergraduate and postgraduate dissertations.

Health and safety assessment of the proposed research is carried out to ensure that the researcher is not being placed under undue risk. Whilst institutions have differing approaches to assessing health and safety and ethics in dissertation research, these are usually addressed in part by the student researcher. The dissertation researcher is usually required to complete a short questionnaire or risk assessment and then sign an accompanying declaration.

With regard to health and safety, the process can involve the carrying out of a risk assessment and scrutiny by specialist academic or technical staff, your academic supervisor, course staff or administrative staff. Health and safety issues may be of concern where research involves fieldwork off-campus or when experimental work is carried out. Your work must not place you or others at undue risk. With regard to ethical considerations, your research proposal will be subject to greater scrutiny if your proposed research involves people or commercial sensitivities. Before approval of a research proposal is granted and the research is able to commence and proceed, the research must have been judged safe and ethical by a member of university staff.

More detailed attention is given to the consideration of health and safety and ethical issues later in the book (see Chapter 7).

The layout of a research proposal is similar to that of the final dissertation (see 'Writing up' in Chapter 9).

The key sections and headings of a research proposal document commonly include:

- Title page
- Summary (or Abstract)

- Contents page
  - Each heading and subheading should have a page number
- Additional resources as appropriate, which may include some or all of the following:
  - List of figures with page numbers
  - List of tables with page numbers
  - List of equations with page numbers
  - List of photographs/plates with page numbers
- Acknowledgements (optional)
- Main body of text:
  - divided into chapters and section headings
- References – sources used *
- Bibliography – sources read but not used *
- Appendices (if required)
  - Ethics review
  - Health and safety review
  - Gantt chart outlining proposed plan of work and progress to date
  - Tables of preliminary results
  - Draft surveys

*You are advised to consult the specific guidance provided by your university or institution as the convention of using the headings 'References' and 'Bibliography' is sometimes the reverse of that shown here.

Instead of an abstract, a summary may be included in the research proposal. It is recommended that this is about 250 to 350 words in length. The abstract or summary should communicate concisely the entire document contents. When composing an abstract or summary, consider including sentences that outline:

- the subject under study (and perhaps why it was selected);
- the methodology proposed;
- key literature and findings to date;
- any conclusions or recommendations which will form part of your research.

The main body of the proposal should detail:

- your aims and objectives or hypothesis;
- key literature regarding your chosen topic of study or identified problem;
- how you propose to collect data;
- how you propose to analyse data.

The research proposal requires the student researcher to focus their thoughts and intentions and document them in an organised manner. This process can be very useful in moving you forward along your research path, even if the proposal results in significant changes having to be made to the dissertation research project proposed initially.

Summary

This chapter has introduced the broad stages involved in doing a dissertation and has discussed issues associated with selecting a topic and title, using a hypothesis and null hypothesis, writing aims and objectives, and bringing together your initial research ideas in the form of research proposal. Student reflections and supervisor guidance were also key components of this chapter.

Suggested further reading

You may find the following references to be of particular use when starting out on your dissertation journey. This suggested further reading identifies books and chapters commonly referred to by the authors in the early stages of supervising a dissertation research project.

Aims

FARRELL, P. (2011) *Writing a Built Environment Dissertation: Practical guidance and examples.* Chichester: Wiley-Blackwell. ISBN 978-1-4051-9851-6

Methodology

DAWSON, C. (2009) *Introduction to Research Methods: A practical guide for anyone undertaking a research project.* 4th ed. Glasgow: Bell & Bain Ltd. ISBN 978-1-84528-367-4
  Chapter 3: How to choose your research methods, pp. 27–39.
GREETHAM, B. (2009) *How to Write Your Undergraduate Dissertation.* Houndmills, Basingstoke: Palgrave Macmillan. ISBN 978-0-230-21875-8
  Chapter 6: Planning your dissertation, pp. 231–57.
MAXWELL, J. A. (2012) *Qualitative Research Design: An interpretive approach.* 3rd ed. Los Angeles: Sage. ISBN 10-1412981190
RUDESTAM, K. E. and NEWTON, R. R. (2007) *Surviving Your Dissertation: A comprehensive guide to content and process.* Los Angeles: Sage. ISBN 978-1-4129-1679-0
  Chapter 5: The methods chapter, pp. 87–115.
SWETNAM, D. (2001) *Writing Your Dissertation: How to plan, prepare and present successful work.* 3rd ed. Oxford: How To Books Ltd. ISBN 1-85703-662-X
  Chapter 4: Techniques, pp. 51–63.

Writing

ALLISON, B. & RACE, P. (2004) *The Student's Guide to Preparing Dissertations and Theses.* 2nd ed. London: Routledge Falmer. 0-415-33486-1 45-61,69-71
BURNETT, J. (2009) *Doing Your Social Science Dissertation.* Sage Study Skills Series. London: Sage. ISBN 978-1-4129-3112-0
  Chapter 1: Ready to do research? pp. 13–35.
  Chapter 4: What kind of researcher are you? pp. 61–77.
CRESWELL, J. W. (2012) *Educational Research: Planning, conducting, and evaluating*

quantitative and qualitative research. 4th ed. Boston MA: Pearson. ISBN 978-0-13-261394-1
  Chapter 2: Identifying a research problem, pp. 59–79.
FISHER, C. *et al.* (2010) *Researching and Writing a Dissertation: A guidebook for business students.* 3rd ed. Harlow, Essex: Pearson Education Limited.
  Introduction, pp. 1–29 (especially the section: Jargon, 'isms' and 'ologies').
GILL, J. & JOHNSON, P. (2010) *Research Methods for Managers.* 4th ed. London: Sage. 978-1847870940
  Chapter 8: Philosophical disputes and management research, pp. 187–213.
  Chapter 9: Conclusions: Evaluating management research, pp. 214–39.
NAOUM, S. G. (2007) *Dissertation Research and Writing for Construction Students.* 2nd ed. Oxford: Butterworth-Heinemann. ISBN 0-7506-8264-7
  Chapter 2: Selecting a topic and writing the dissertation proposal, pp. 11–17.
  Appendix 1: Examples of dissertation proposals, pp. 171–86.

4   The role of the supervisor

Introduction

This chapter outlines:

- the general role of a supervisor and what a student might expect from them;
- a few of the issues which may occur between students and supervisors, and how to manage these;
- things to consider in order to effectively control the process of undertaking a dissertation;
- virtual supervisors' views on supervision;
- students' own experience of the supervision process;
- suggested reading.

Managing the student–supervisor relationship

This section has been deliberately titled to counteract the most common perception from students: that a supervisor will always dominate and direct the dissertation process. In some cases, this may be true; but in the majority, the student is the driving force behind the delivery of the project and will need to act in such a way as to optimise the input from their supervisor. The key to success for a student may be careful management of both their supervisor and the research.

As a general rule, a supervisor will:

- know what is expected of an undergraduate dissertation and be able to provide guidance that helps a student achieve a pass level;
- be able to provide a student with a meeting (face-to-face, over the phone, via Skype) to provide formal supervision or, alternatively, to provide help and guidance by email;
- have knowledge of some of the techniques available for research;
- help a student to schedule the work for deadlines;
- be able to direct a student towards potential sources of available literature;

- suggest specific techniques of data acquisition or analysis;
- provide some feedback on written work.

The student needs to:

- find out what is expected of them;
- negotiate format, number and length of meetings with their supervisor;
- arrange times and dates of meetings;
- set meeting agendas;
- attend meetings, negotiate and set targets for review at the next meeting;
- bring work and questions to each meeting;
- provide reports on progress and updates;
- list problems or areas of concern or uncertainty;
- inform the supervisor of problems or difficulties in a timely manner, as they occur;
- take notes during the meetings and produce action points based on the decisions made;
- undertake the work agreed.

The supervisory process is most often based on human interactions and discussions. Any supervisor is usually working with the best interests of their student(s) at heart. Just like students, supervisors are subject to aspects of the human condition, including but not limited to: sickness; family illness or bereavement; having other conflicting demands on their time; and making mistakes. Students should be aware of the importance of dialogue and of managing and maintaining effective communication with their supervisor throughout the process.

Understanding differences in supervisors – some generalised descriptions

It is highly unlikely that any supervisor will start with the deliberate intention of working against a student to ensure that the work fails. Supervisors are there to support and encourage students through a challenging piece of work. However, not all people are alike; supervisors, like students, work in different ways. Supervisors have different personalities, motivations, strengths and weaknesses; they also have differing goals and priorities, various levels of education or experience, diverse backgrounds and preferences, and dissimilar views on teaching, learning and research. Understanding that these differences exist will help a student to manage their side of the process more effectively.

The supervisor guidance on the role of the supervisor gives an insight into how their views may differ. The generalisations in the box on styles of supervision represent some of the approaches that may be taken. These may also change with time; for example, a supervisor with a 'hands-off' approach may become increasingly controlling if it becomes apparent that the student work is going badly and risks failing.

Styles of supervision

The 'what are you going to do and how?' supervisor

Style　　　Most of the supervision time is spent at the beginning of the project and relatively little on the final stages. This type of supervisor will ask lots of questions which are generally focused on the student's intended methodology.

Advantages　The student is in control of the topic and how it develops. As long as the student process is sound, the supervisor will be happy.

Disadvantages　The lack of supervisory input into the final stages can lead to student anxiety.

Management suggestions　Students should book lots of early meetings to get the full benefit in terms of refinement of the process. Clear notes should be made about how the work is to be structured. Ensure that meetings for later in the process are also booked at these early stages so that advice is available throughout the process.

The 'this may reflect on me' supervisor

Style　　　This supervisor treats the work as their own personal project, and efforts are made to ensure that the student is firmly controlled.

Advantages　For most students, this is a comfortable and familiar arrangement. The supervisor is personally interested in the work and may wish to see regular updates and keep a close eye on the work as it proceeds. The supervisor is likely to be proactive in providing suggestions and comments.

Disadvantages　The student does not get an opportunity to fully develop their skills and to challenge the supervisor's view on the topic. The work is likely to be forced into one particular methodology, and the student may not be comfortable with this.

Management suggestions　Students are advised to keep up to date with the work; to share often and in detail; and to be keen and interested. If the student feels that the work is beginning to travel into an area in which they are not comfortable, this should be discussed and negotiated with the supervisor early on.

The 'you're in charge' supervisor

Style The supervisor takes the view that the student is in charge of the work. They may have multiple other internal or external commitments, and booking meetings with them can be challenging. Students are expected to manage the arrangement of meetings.

Advantages The supervisor may sometimes offer remote supervision, which can be more time-effective, and they may give good-quality feedback.

Disadvantages Students may feel that if a supervisor is not answering emails/phone messages or meeting with them then they are not being supervised.

Management suggestions If a student is not able to arrange meetings to see their supervisor then this needs to be raised as an issue with someone who can help in a timely manner but without causing unnecessary early conflicts that will damage the relationship. It is advised that records of requests for meetings that have not been answered are kept. If the tutor doesn't suggest dates and times for meetings, offer multiple options. Keep pushing.

If work is sent by email for comment then non-replies should be followed up. Do not assume that no reply is the equivalent of no issues with the work. If there is still no answer after a few days, begin to chase to ensure that the supervisor is not ill or on an extended leave of absence.

The 'disengaged' supervisor

Style The supervisor does not appear to care about the student work and will give no firm views on progress or content. Directions are generic.

Advantages A disengaged supervisor offers no advantages to the progress or outcomes of the research.

Disadvantages The student may feel isolated and be lacking in necessary guidance.

Management suggestions Write up questions to take to each meeting, annotate with the responses given to make sure a full record of the advice given (or lack of it) is available. Send copies to the supervisor. Stay focused on delivering a good piece of work. As a final resort, request a supervisory change.

The 'my way is the only way' supervisor

Style	This supervisor demands that work must be done exactly as they say, when they say it, and how they dictate it must be done. Meetings may quickly become a supervisor monologue comprised of 'to do' lists for the student.
Advantages	A student is very firmly guided into tried and tested research methods. Delivery of the expected work should result in a pass.
Disadvantages	This is not really the student's own work as all of the creative thinking is done by the supervisor. The student risks becoming disengaged as they are not allowed to disagree with their supervisor. In addition, if the supervisor's demands are not met then they may disengage from the process.
Management suggestions	Use the advice from the supervisor. Meet regularly and discuss any deviations from their advice in terms of focus (perhaps make intimations that it was due to their advice that reflection on the work suggested a parallel research path). This will sound out supervisor views before changes are committed to.

The 'only positive comments' supervisor

Style	This supervisor does not want to upset students and will go out of their way to lavish praise on any and all work, however mediocre.
Advantages	It is always nice to be praised on work, and meetings are likely to be very positive and friendly.
Disadvantages	Sometimes it is necessary to hear uncomfortable truths about work. Critical feedback in the short term may be hard to take, but it is generally made for the long-term good.
Management suggestions	The student should be honest with themselves and should reflect on whether the work really is as good as they are being told or whether they can improve it. Students should request feedback in terms of 'what can I do better?' In that way, the initiative for improvement comes from the student.

Setting the agenda for meetings

There is no fixed approach to planning, doing and recording meetings. The stages addressed throughout will vary. The number of meetings and their format will vary between institutions and, within an institution, between disciplines and individual staff. Tables 4.1 and 4.2 give two suggested outlines that reflect some of the common discussion points covered in research supervision meetings, according to a a six-meeting format and a five-meeting format.

Table 4.1 A draft agenda for a six-meeting format

Meeting	Discussion points
Meeting 1	• The role of the supervisor • The methods by which supervision will take place (face-to-face meetings; email; FaceTime/Skype) • Stages of the research process • Research topic • Aims and objectives • Dissertation assessment criteria
Meeting 2	• Progress made and problems encountered • Review/refinement of aims and objectives • 'Trawl of literature' – what has the student read and noted to date? • Possible key texts • Developing and determining methods for data collection and data analysis • Research concepts – Q&A and discussion of 'research traditions'
Meeting 3	• Outline of proposed research model, ethics, and health and safety matters • Research concepts – Q&A • Primary data collection – discussion of design and management issues • Secondary data collection – progress review
Meeting 4	• Progress review – progress made and problems encountered • Review of data collection • Data analysis
Meeting 5	• Data analysis and discussion of findings • Progress review and key student issues • Presentation of the thesis
Meeting 6	• Findings • Possible further research • Limitations of the research • Presentation of the thesis • Discussion and Q&A regarding the *viva*

Table 4.2 A draft agenda for a five-meeting format

Meeting	Discussion points
Meeting 1	• Initial meeting to discuss proposed title and proposed research • Discussions around literature • Discussions around methods • Targets set for next meeting
Meeting 2	• Review previous targets • Ethical or health and safety implications of methods discussed/formalised • Ongoing literature review • Development of methods • Targets set for next meeting
Meeting 3	• Review previous targets • Progress of research

Table 4.2 (Continued)

Meeting	Discussion points
	• Data collected and analysis intended
	• Targets set for next meeting
Meeting 4	• Review previous targets
	• Progress of research
	• Initial findings and ongoing analysis
	• Targets set for next meeting
Meeting 5	• Review previous targets
	• Write up commenced – meeting reviews progress and suggests any improvements
	• Targets set for handing in of work

Dissertation progress checklist

An alternative to the simple target-based approach is to draft a more comprehensive checklist for each meeting. Over the course of the entire supervision, it would be expected that all of the following aspects are discussed, with the focus changing over time:

1. *Purpose of the meeting* – To update the supervisor on the progress being made on the dissertation to date and the plans for completing the project.
2. *Aims and objectives* – Discuss with the supervisor the aims and objectives for the research. (These may have changed as the research has progressed and developed.)
3. *Hypotheses or research questions* – Ensure that the hypotheses or research questions are realistic and achievable and that there is agreement between student and supervisor.
4. *Literature* – Discuss the key issues that have come out of the literature review.
5. *Methodology* – Outline in broad terms how data may be collected (the 'instruments' to be used) and from where. Sampling selection should also be considered in general terms at the early stages.
6. *Ethics* – Ethical consideration of factors involved in the research is presented. This must be done even if there is a low possibility of harm. There should be a record that the supervisor has reviewed the research ethics.
7. *Risk assessment* – Discussion of any potential health and safety risks with your supervisor.
8. *Data analysis* – Early conversation on how it is proposed that data are analysed. This may be linked to approaches identified within the literature.
9. *Timescale* – A draft timeline for completion of the work with key targets identified. A schedule of meetings should also be agreed to monitor progress. Care should be taken to identify any points in the calendar where student or supervisor availability is limited (for example, over holiday or exam periods).

Whichever format is followed, it is important that the meeting is recorded clearly as these notes may be pertinent several months after the meeting. An example record sheet is given in Figure 4.1.

RESEARCH SUPERVISION RECORD

Name of candidate:

Name of supervisor:

Date/time of meeting:

Meeting no:

Summary of matters discussed/feedback on progress

Items for progressing/feedback and feed forwards

Date/time of next meeting: ....

Supervisor's signature: ....

Candidate's signature: ....

Figure 4.1 Research supervision record form

Researcher question
Managing the student–supervisor relationship

Do you have a broad understanding of the nature of the student–supervisor relationship and how challenges may be addressed?

Yes. I am aware of the broad expectations on me to manage the process with supervisor input.

No. I am still unsure about this. You are advised to reread this chapter and identify strategies for management. Any further uncertainties you have should be discussed in person with your supervisor.

Supervisor guidance on the role of the supervisor

Guidance from Supervisor A

The role of the supervisor is something that I discuss during the first supervisory session with each dissertation student, regardless of whether the session is face-to-face or via video conferencing, telephone or email. My personal preference is always for the first supervisory session to be either a face-to-face meeting or a video conference chat.

In my experience, supervisors and students alike can have greatly varying views about the exact role of the supervisor. Some see the role as being akin to that of a doctor, dentist or car mechanic – a visit is made either a) periodically to 'check things over' or b) when things have gone wrong.

I am sure that there are many comparable metaphors that could be used to describe the possible role of the supervisor. For me, the supervisor's role is one that consists of a mix of listening, suggesting specific reading, answering questions, discussing research management issues, giving informed views and opinions, and encouraging progress. Supervisors are people with previous research and supervisor experience.

Please do bear in mind that a research supervisor cannot and will not do the research for you; they are not your own personal 'research assistant'. The role of the supervisor is one of being a 'critical research friend': someone who will answer research-related questions and who will support, guide, encourage and possibly inspire you.

Guidance from Supervisor B

The role of a supervisor is very personal and varies considerably due to the relationship between the student and the supervisor as well as the topic of research and methods of investigation chosen. I'm not sure that students really understand this at the beginning of their journey, and perhaps they expect their supervisor to be more like a tour guide, pointing them to key resources as they move from a defined start to a defined finish. It really isn't always like that. Some supervisors will do that certainly; others will be more like a guidebook or map that you consult when you need directions or more information. Others will behave as if they are accompanying you and asking you questions about what you can see and what that means. Some supervisors are very strict with regards to meetings and want to see you with progress on a regular basis. This can work for some students, but it may seem like being hounded to others. At the other end of the spectrum, there are supervisors who will only want to see firm results. Most supervisors will not offer detailed proofreading of your work, but many will overview parts of it for style or content or discuss your results as you collect and analyse your data. You, as a student, have a great influence on the relationship and the role taken by your supervisor; and if you communicate your needs (and these are reasonable) then your supervisor can be your greatest ally in completing your work.

Guidance from Supervisor C

A supervisor is there to ensure that there is firm guidance and clear academic rigour to the work produced by the student. Supervisors are not there to assist in the interpretation of data, although of course I enjoy discussing student findings. My role in the process is to provide expert guidance, signposting the way to approach the topic clearly so that my students gain the greatest experience from the process. I enjoy sharing the challenge of discovery and seeing a student become an autonomous individual. I want my students to fully engage with the whole process and deliver a piece of work of which we are both proud. This means that I do not spend my time attempting to manage meetings. At this stage, the student is fully expected to request my time as and when they need, within the bounds of reasonableness. My remit is to get them from start to finish, enabling them to achieve the best grade they possibly can by delivering a sound piece of academic work.

Student reflections on the role of the supervisor

'My tutor was second to none, offered detailed advice and guidance and responded quickly to any queries, and allowed me to have as many meetings as I liked. He was great, really helpful, approachable and enthusiastic about the topic. He also encouraged me to look deeper and challenge my level of involvement to achieve a deeper one.'

'The main problem I had as a student was that I was expecting my supervisor to have all the answers. I would go to meetings and my supervisor would ask me what I thought about a particular issue, and I was just thinking "well you already know this so why aren't you telling me?" I found that part really frustrating. I just didn't understand why, if your supervisor was supposed to be an expert in the field, they weren't just telling you what you need to know to do the project. It wasn't until much later that I realised that the whole point was that I was supposed to be finding these things out and then talking to my supervisor, rather than just accepting that he had all the answers.'

'To begin with, I felt the relationship with me and my supervisor was not as good as I hoped. I sent emails but they were not being replied to. After a month or so, I realised the best form of communication was simply to phone. This worked brilliantly, the relationship was much better, and I got all the help I required.'

'My supervisor was a bit hard to contact and get hold of at times when arranging meetings, especially early on. However, once a meeting was sorted he was great.'

'My supervisor was ill, but I have found that the module leader answered all of my questions, pointed me in the right direction, encouraged me and provided constructive criticism and feedback – all of this done promptly at my request.'

'My dissertation really didn't go very well. I didn't realise my supervisor was on sick leave. I sent work by email requesting feedback if there was a problem and there was no reply. Although I was frustrated that my emails were not answered, I assumed that my work was OK and I didn't try and ask anyone else. In hindsight, it would probably have been sensible to phone up and find out what was happening rather than just ranting to friends that my emails hadn't been answered.'

'It really is up to the student, so if you need assistance and you ask for it then I'm sure the supervisor will oblige. On reflection, perhaps I didn't use the supervisor to their full potential; however, they were happy to let me crack on as I saw fit. I quite liked having my independence to carry out my own research project.'

'I feel that working to this level should not be about hitting a matrix-based marking scheme. It should be about going out and producing something that is yours. It should be "free roaming". It irritates me when you see people constantly with lecturers looking for coursework hints and advice. This doesn't happen in industry, so it shouldn't happen here! By this point of our university lives, it should be down to the individual's ability, not the tutor's.'

'My tutor was inspiring all through the process, and I really felt that I was well supported. I felt able to ask questions (even though some of them felt a bit silly). She was always on hand to help and gave rapid and helpful responses. It was nice to know that I wasn't on my own with the feeling of panic.'

'From talking with my peers, all tutors appear to do things extremely differently.'

'I was very satisfied with my supervisor. It was really helpful to focus on my topic and to receive guidance on collecting primary and secondary data. Students might have the idea; however, the focus seems unclear. My supervisor has helped me to outline the important elements and objectives.'

'I was fortunate enough to get a supervisor I knew from previous modules who I always found to be an excellent tutor with great work ethics. Having him as a supervisor on my dissertation was an absolute pleasure; he made me think outside the box and offered advice in times when I felt lost along the way. He made me think logically about what I was doing and made me think about things differently to what I would usually. This helped me no end during the whole process.'

'I thought my supervisor was great – not critical to the point of making me cry but very constructive and supportive in her guidance. She didn't insist on meetings when not needed but was there when I needed some guidance. She was interested in my topic and gave sound, practical and honest advice. I never walked away from a meeting thinking it was a waste of time; I always had something that I could take away to improve on or work towards. And often I felt better after a meeting as it enabled me to break the dissertation down into manageable chunks as opposed to trying to look at it as a whole.'

'I had a good relationship with my tutor but it wasn't as helpful as I thought it would have been. Initially I expected that he would be pushing me to complete each section, possibly offering some insights of his own from his experience; but I felt that I was independently undertaking the research with very little assistance. While at times a helping hand would have been greatly appreciated, I came to the understanding that this was a final-year module and so I was expected to work independently.'

Summary

This chapter has introduced the role of the supervisor and stressed that the effectiveness of the student–supervisor relationship can depend strongly on communication. The highly individual nature of the experience has been described, and some techniques for managing the process have been suggested. Recommendations were made for supervisory meeting agendas and means by which students may monitor and report their own progress.

Suggested further reading

FARRELL, P. (2011) *Writing a Built Environment Dissertation: Practical guidance and examples*. Chichester: Wiley-Blackwell. ISBN 978-1-4051-9851-6
  Section 1.6: The student/supervisor relationship and time management, pp. 9–10.
NAOUM, S. G. (2007) *Dissertation Research and Writing for Construction Students*. 2nd ed. Oxford: Butterworth-Heinemann. ISBN 0-7506-8264-7
  Chapter 10: Dissertation supervision and assessment, pp. 161–70.

5   Evaluating the existing literature

Introduction

This chapter outlines:

- the nature and types of secondary and primary sources of data that can be used in student research;
- the nature of the literature review, how this may relate to student work and how to build in references to published work to show the linkages to the existing body of knowledge;
- the clear use of citation and referencing and how to undertake this within the written work
- guidance from virtual supervisors;
- student reflections on using secondary data;
- suggestions for further reading.

Secondary approaches

There are, fundamentally, two sources of research data (information): primary data and secondary data.

Primary data is information collected explicitly for the research. Although it is not always appropriate to a piece of research, most students seem to think that they must include some primary data in their dissertation.

Secondary data is information that has been generated by an individual other than the user. This usually comprises written works to which a student should refer in order to understand the existing body of knowledge. A dissertation will invariably include such data. Secondary data may also include numbers, photographs, statistics or any other data that has been collected by another individual for a purpose other than the dissertation research.

At the BSc (Hons) degree level of research, a student can be involved with primary and secondary data collection. A dissertation may include both primary and secondary data or (less commonly) only secondary information. A summary of advantages and disadvantages is given in Table 5.1.

Table 5.1 Advantages and disadvantages of primary and secondary data

Source of data	Advantages	Disadvantages
Secondary	Already exists, and may have been analysed Researcher is more in control of work flow Easy to collect a large amount of data	Data may be only marginally related to project A wide amount of reading is needed Data can be outdated Data need to be interpreted by researcher
Primary	Gathered by the researcher Researcher has control over volume and nature of data Many options available Material is up to date	Data collection may be time-consuming Collection needs careful planning Circumstances may result in suboptimal data set Data need to be analysed

Research approaches that collect secondary data

Sources of secondary data may include company records, maps, photographs or census data. It is often data that would be impossible for a student to collect on their own due to the scale of the collection or the historical nature of the data.

The term is also commonly used for books and research papers which may contain a combination of primary data collected by the author of that work and secondary data that they have used to compile the work. Secondary works may include published case studies (these are not to be confused with case studies done by a student to collect primary data); published interviews (done by another); any literature (books, magazines, newspapers, online articles, news articles); audio-visual sources such as documentaries; or statistics, processed and published. In some cases, only secondary data are available, particularly for archival research.

Where the literature review is designed as part of a research inquiry, there will be a more logical connection between the literature review and the student's own work. Secondary data from existing research can be useful in suggesting methods that may be appropriate to student research projects.

For a dissertation or research project based around a design-related inquiry, the literature will underpin development and will not necessarily be directly connected with the design object.

Using secondary sources has considerable advantages in terms of time needed to find information. The resources are often readily available and easy to access. A common disadvantage is that there is too much information available which can leave a student feeling swamped. Alternatively, access to some sources may be difficult, needing inter-library loans or personal access to the place holding the resource, which lends uncertainly to the process.

The main research methods based on information of a secondary type are:

- annotated bibliography;
- literature review;
- historical data analysis.

An annotated bibliography is an intermediate point towards a full literature review. In this, the literature is often presented as a series of paragraphs. The student provides a summary of the work in terms of key findings and makes a response to the work that shows understanding of the significance and relevance to the underlying topic area. This is often organised into themes and is then used to produce a full literature review.

Historical data analysis may be based on a variety of sources. Sources from archives may include official documents or other items stored in a collection, which could be letters, personal recollections (see below) or historical photographs. These could be held in an institution (e.g. a museum or public archive) or privately (in a personal collection or private archive).

In addition, records taken over a period of time (including estate accounts, statistics, census data, and property deeds) may be found within archives or in libraries. Data of this type are the sort most commonly used in historical research. Other possible sources include written recollections of the events of a particular time, which may be in the form of memoirs or diaries that have been collated and published. An example would be Ruskin's work on the historical aspects of the architecture of Venice.

Historical data are often incomplete and biased in nature, and this complexity can lead to multiple and conflicting reinterpretations found in different secondary sources, sometimes based on identical primary data.

The literature review

A literature review (also called 'literature research study' or 'literature survey') groups relevant ideas and evidence from secondary sources into a logical flow, addressing significance and understanding. This demonstrates that the work is built soundly on pre-existing ideas. It can also be used to show where existing work has been used in a new way within the student's work.

An alphabetical list of the secondary sources used is usually presented separately. It is a recognised and accepted part of academic research projects at all degree levels. All research, regardless of scale, requires a certain amount of reading about: what other people have written about the area of interest; and how to collect information for the topic being researched, including what methods have previously been used by others. For an undergraduate dissertation, the expectation is not that the student will produce a definitive and comprehensive account of all sources. However, the student will be required to demonstrate that they have considered an appropriate amount of the relevant literature and that this literature has been derived from suitable, predominantly academic sources.

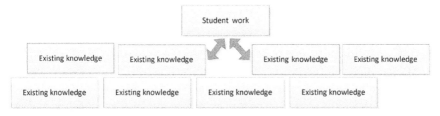

Figure 5.1 Model of the relationship of dissertation work to literature

Figure 5.1 shows a model of how student work builds on existing work to bridge gaps in that knowledge. The student work repeats or builds on some aspects of the literature. An understanding of the literature helps with formulation of student aims and objectives by demonstrating what has already been done and how. Indeed, the researcher has the option to explore a very limited aspect not previously covered in the literature. Here, the literature can be seen to be vital to supporting and underpinning the student work. Note that not all topics necessarily require such gaps to be identified. The literature might be used instead to place the student's findings in the context of existing findings.

It is suggested that most of the reading should be done early, even before the 'official' start of the dissertation and certainly within the first three to four months. Relevant material might be in the form of:

- textbooks (paper or e-books);
- journals (paper or e-journals);
- research papers (paper or electronic format);
- conference papers (paper or electronic format);
- broadcast information (video, radio, CD);
- company reports;
- market research;
- digests of information or databases, abstracts or indexes;
- business directories;
- World Wide Web (provenance should be considered);
- previous dissertations or theses.

Reading will usually continue throughout the period when data are being collected and, in some cases, as late as the data analysis stage. As a caution, reading time should be carefully controlled in these later stages as much time can be taken up for relatively little credit.

References should be written up in Harvard format (or whatever format is specified by your home institute) as work progresses.

A thorough literature review is time-consuming but worth doing right first time to prevent wasted time in the writing-up phase. There are no shortcuts, but the following ideas may help:

- identify the sources of information which are accessible;

- become familiar with your library search engine as well as Google search, and use bibliographic databases;
- clarify the topic and generate appropriate keyword searches for the main concepts;
- be flexible;
- there will often be several ways to describe a topic and synonyms (alternative words) should be used to make sure as much material as possible is traced (a good thesaurus may be of help in finding synonyms);
- keywords can be used to search the major databases in the chosen subject area(s) for journal articles and books that will help you with your research;
- use automated alerts for keywords.

You MUST keep a full and accurate record of all the materials used.

Examples of the literature review

The collection of sources selected for inclusion in a literature review may illustrate agreement or contrary views relating to a topic. The sources may highlight areas where little work has been done to date or aspects where there is a lot of existing work. A review should provide the reader with a concise summary of the current state of knowledge to allow them to see how the student's research relates to and builds on this. The examples that follow illustrate different levels of practice.

Literature review example 1

This process of good communication and information between the Architect and Project Management does not end there. It also has to take place on-site and be given to the workforce.

Goodman & Love (1980, p36) state *'Thus far, management tasks have been categorised and discussed in two ways: 1) by general function – planning, organisation, staffing, directing, and controlling supported by an information system, a communications system, and a responsibility/authority system; and 2) by task of project cycle – project identification, design, appraisal, selection, organisation, operation, supervision and control, completion, and continuation.'*

Stallworthy & Kharbanda (1985, p85) state: *'To get things done, the project manager has to ensure a constant and prompt flow of information and instructions through his project team to the entire workforce – a workforce that can range from a few hundred to several thousand on a large-scale project. Effective communication, therefore, whether written or verbal, is absolutely essential.'*

Lock (1994, p5) states: *'Project management techniques find their application in many situations far removed from the obvious industrial project scene, helping to manage changes of premises, new installations, refurbishment or maintenance of existing plant and facilities, company relocations and so on'* and *'Every project manager must ensure that his project is efficiently and sensibly planned and scheduled from the start, and that critical tasks are identified'* (Lock 1994, p6).

Critique

Only 30 words of this 217-word section are original words written by the student. The rest is assembled quotes with no evidence that the student has understood or discussed the views expressed.

It is clearly referenced, although many references are quite old.

Overall assessment

Needs improvement

Literature review example 2

As economic demands and higher student expectations relating to quality are placing more strains upon educational institutions, there is an increasing need for a holistic way of meeting these sometimes diverse demands. Few are willing to define quality in Colleges and Universities, although many claim to know intrinsically when it is present. Similarly, faculty and administrators alike are reluctant to call a student or anyone else a customer. This environment presents strong objections to the language, principles and methods of Total Quality Management (TQM). The principles of TQM are neither new nor unique.

Wasson (1993, p. 109) advocated that the requirements of TQM were *'simply good management codified'*. However, TQM's applicability to education has been questioned. Its critics state that 'customer' is an inappropriate term. For TQM to work *'... educational establishments must focus on the students as the customers'* (Harris 1993, p. 17).

Therefore TQM could assist educationalists in viewing their actions from a perspective that has the *'customer/student'* at the centre of its organisational activities. Liebmann has implemented TQM within an educational establishment and advocated that *'the basic TQM theories did seem sound'* (Liebmann 1993, p. 18).

Mooney opined that: *'TQM makes the customer the focus of all decision making. In educational terms the external customers are the students and the parents and the product is learning. Teachers are "service providers" providing the service of teaching'* (Mooney 1993, p. 151).

Howard, who also implemented TQM within an educational environ-ment, found that the '*key to effective TQM is to develop a position aimed at meeting our unique institutional needs in a manner that is consistent with our organisational culture and philosophy*' (Howard 1993, p. 110). Further support for the deployment of TQM within education comes from Lozies and Teeter (193, p. 11). '*The TQM foundations for the pursuit of quality can have a powerful impact on efforts to improve higher education.*' As a final point appertaining to the avocation of TQM within education, Sutcliffe and Pollock (1992, p. 26) stated '*You don't have to do this – survival is not compulsory*'.

Critique

Quotes make up 114 of the 339 words with the remainder, original student work. Several sources are clearly referenced although these are relatively old and more recent sources should be available. Paraphrasing used (authors own words using the ideas of others) with referencing. Shows discussion.

Overall assessment

An improvement over example review 1

Literature review example 3

The Lambert review (2003:7) showed that companies are broadly satisfied with the graduates which they recruit, although there was a mismatch between supply and demand of graduates in some areas. Conversely Handel (2003:135,149) stressed the common belief of a mismatch between educa-tion and the required skills of the workforce within the current economy, also reporting a dearth of information about '*what people actually do at work*'. Increased employability is often underpinned by the provision of work-based learning activities, the impediment of employability in the course design and the creation of academic and careers adviser partnerships (Pegg et al., 2012:11). Purcell et al. (2012:64) suggest that recruiters continue to state that there is an inability to recruit appropriate talent in particu-lar graduate vacancies, and yet many graduates still report that they are unable to find graduate vacancies and thus there is a net underutilisation of skills which will become increasingly important in light of increasing graduate debt.

Critique

Paragraph starts with a citation and, in general, it is advised to put infor-mation first and citation at the end of the sentence. Quotes make up

6 of the 157 words with the remainder, original student work. Several recent sources are clearly referenced and illustrate contradicting views. Paraphrasing used with referencing.

Overall assessment

An improvement over example review 2

Writing a literature review

The literature review should be informed and critical and demonstrate an understanding of the topic, the approaches and theories, the arguments or facts. When writing the review, it is suggested that an introductory paragraph, which clearly scopes the topic area to be covered, is given. The main body of the text should then:

- Identify trends, themes, conflicts and key authors or sources.
- Present a balanced view or use a balanced and authoritative range of sources suitable for the work. Good-quality journals or textbooks should be considered the default. In some cases, newspapers or Internet sources may be appropriate in addition to but rarely to the exclusion of peer-reviewed sources.
- Provide the latest view on the subject. (It is suggested that students should avoid the temptation to only recycle sources used in dissertation theses from four or five years ago as they are not likely to be the most relevant sources available.)

Consider using an outline to ensure that the literature review links clearly to the work to be delivered. Subsections may be used to increase clarity for the reader.

Finally, there should be a summary of the literature review that identifies the areas which are:

- important to the work;
- linked (where one area of previous work links to another area of previous work);
- to be verified;
- subject to change(s);
- conflicting;
- missing;
- linked to research questions;
- based on methodologies similar to the work being done (which could help to justify dissertation research choices).

Researcher question
Writing a literature review

Do you feel that you have a working understanding of the purpose and format of the literature review?

Yes. I understand this aspect of the work and feel confident that I can identify and use appropriate sources of literature within my work.

No. There are aspects of either finding resources or being able to use these correctly that I am not sure about. I recognise that I will need to undertake further reading or gain additional support, which may need to be specific to my needs. I undertake to approach these sources of help which may include my home institution's library, my supervisor or further reading.

Theory and literature

An essential early stage of virtually all research is to search for and to examine potentially relevant theory in the literature. Some sources may be based solely on discussions around theory, but theories can also result from or be refined by previous research projects. This may include research findings that have not attained the status of theory (principles and laws); often it represents findings from research into particular applications of theory.

The relevant theories and major references related to these should be established in early discussion with the supervisor and others who are experts in the topic. Consultations to determine the usefulness of the proposal during its formulation will reveal appropriate theories as well as major research projects that have been carried out in relation to these. These are good starting places.

Analytical frameworks

Reading about the topic as much as time permits may allow ideas about approach and methods to be formulated which would otherwise not have been considered. Literature may contain examples of how data can be classified and presented, or it may help in the derivation of a theoretical or analytical framework for interpretation and examination of the data. The facts collected have to be arranged, classified and formed into a coherent pattern in order to allow a full explanation of the relationships between facts as observed.

Using the literature to see how other authors have collected the data, analysed their findings and generated relationships can be useful when undertaking the dissertation research. The processes used vary widely between disciplines and some methods may be seen as more valid than others when dealing with certain types of data or particular research traditions. However, knowing that there are alternatives can be useful when arguing for the applicability of the chosen method and justifying the methods used to reach the conclusions.

Planning a literature search

The key to a successful literature review is in preparation and recording. Detailed notes and clear references should be kept and reviewed to ensure that time is well spent. It is suggested that a summary of the key points is made while the material is fresh in the student's mind.

The following outlines general good practice:

- Search thoroughly (do not risk missing a really obvious, relevant and up-to-date source).
- Keep reading to ensure continuous awareness and appraisal of any changes in the field of work.
- Write early and update regularly. This 'forces' continuous engagement and thought about the topic. Writing up has to be done at some stage in any case. It is advised to start early and work on an ongoing basis rather leaving it until late and discovering that the work done has missed something fundamental.
- Reference clearly. Keep reference lists up to date.
- Keep a diary and to-do list for future ideas handy.

Locating published materials

If a topic has been provided for investigation, then in all probability a list of books and articles will be available from your supervisor which should be listed as 'required reading', 'recommended reading', 'suggested reading', 'set books', or similar. Where provided, these sources should always be seen as a starting point, not a finish point. If the project is small (two to three months from start to completion) then these may form most of the literature needed for the topic. If the project is longer or in the absence of such lists then it is for the student, with or without supervisor input, to locate and use published sources.

A student therefore needs to:

- find the most relevant published materials quickly;
- avoid getting mired in irrelevant sources;
- get into good habits early with regard to recording information so that this can be easily found and understood weeks or months later.

Whatever the size of the task, meticulous planning and attention to detail needs to be adopted in conducting a literature search.

Students should ensure an understanding of the services offered by their university library and library staff (ask at the library information desk). It is useful to become familiar with what materials the university library holds or has access to either physically or virtually; equally, it can save time to find out what materials are not held by the university library and where these can be obtained. All libraries vary to some extent, so some time needs to be invested to become familiar with the geography and the stock before the search for literature begins. This need extends to the electronically accessed resources available through the library catalogue.

Access to other libraries

PUBLIC LIBRARIES

Run by the local authority, the size of public libraries and the services on offer will vary considerably. Students living locally may be able to borrow resources; if not, it may be possible for a student to do research work on the premises.

ACADEMIC LIBRARIES AT OTHER UNIVERSITIES

Students from other institutions who wish to use the facilities of an academic library will need to obtain permission. SCONUL Access is a scheme that allows students at participating universities to use the facilities of other libraries which also belong to the scheme. Using the SCONUL website, it is possible to find the contact point to request access. Other university libraries will allow the public access although this is sometimes at restricted times, for example during student vacation periods.

Focusing a search

Many a research project has foundered because the investigator did not define the area of study sufficiently clearly and so extended the range of reading far beyond that which was necessary. Figure 5.2 illustrates the process of refining and focusing in on the most relevant sources over the course of the review.

Large libraries are complicated places and it is easy to become lost and frustrated.

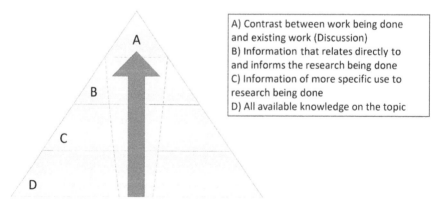

Figure 5.2 Diagram showing the progression of a literature review

Table 5.2 Planning the literature review

Stages	Notes
1. Select the topic	This should be in the title of the work.
2. Define the terminology	Ensure that the primary terms selected in stage 1 are sufficient to find all literature from this area. Consider that, for example, different terms or spellings may be used in some countries.
3. Define the parameters	Consider from what sources the literature will be drawn: • geographic area (e.g. UK, Europe, America, Africa, Australia) • time period (e.g. last 5 years, 10 years, 100 years) • type of material (books, journals, theses, reports) • sector (some literature may be very specific to one industry or sphere of practice).
4. List possible search terms	Synonyms of the word will provide alternative search terms.
5. Refine	As the search progresses, other search terms will become apparent. Repeat from stage 1.

Library staff at the information desk will do their best to help you come to grips with the way the stock is organised. It is necessary to be clear about what exactly is being looked for, and it is only possible to get to know what a library can offer and how to use the facilities when the information search begins. Electronic access to journals has greatly increased the range of readily accessible information. It is easy to be swamped without careful consideration of the principal focus of the research. Table 5.2 suggests a way to think about the process of planning.

The literature review as the sole data collection mechanism

It is often asked if a literature review on its own constitutes a sufficient level of investigation for a dissertation. It may well be sufficient subject to a number of provisos:

• that the literature review is very comprehensive;
• that the literature review collects information from all principal sources.

Researcher question
Planning a literature search

Do you feel able to locate appropriate literature-based sources?

Yes. I have undertaken preliminary searches on my proposed topic area and have been able to find appropriate literature sources

No. I have not yet undertaken a preliminary review of my topic area. I understand that I will need to be able to identify appropriate sources and may need to seek additional help if I am struggling to use the systems effectively.

The key aspect to utilising a literature review as the sole source of data is that the literature review *presents clear evidence of critique*.

If this methodological approach is to be adopted then it is essential that the supervisor establishes clearly, with the student, what is involved from the outset. A more comprehensive and detailed literature review should be expected from a student who adopts this approach. The approach is not quite as simple as it may appear because the student may be involved in searching for information from unusual sources and using unfamiliar methods. In addition, the approach could be more demanding on the supervisor who may need to verify the, perhaps, many and different sources of information used in the work.

The element of critical appraisal is essential within this approach and there must be clear evidence of it. A student *must* ask the supervisor if literature review alone is suitable. A critique should be a balanced view of the issues or problems investigated within a subject area. This aspect is one that undergraduate students can find hard to grasp, in particular when writing up such work. It can also require that the supervisor knows more about the subject matter in order to take an equally balanced view.

A further issue in using a literature review as the sole investigative method for research is the ability of the student to take a view of the work in the wider context. Certainly, the broader the data (literature) base, the higher the expectation of seeing the work in the broader context.

Historical data

Historical research may involve investigating complex phenomena of social and physical interactions that need to be explained and interpreted. Historical research is commonly qualitative in nature, and the researcher should try and collect as much evidence as practicably possible in order to provide an appropriate evidence base. In addition to the traditional literature review, primary evidence in the form of manuscripts, unpublished records, company documents, or personal letters may be used.

In addition to these written sources, 'artefacts' such as buildings, photographs or other tangible sources may form the study materials. For instance, the study of architecture as built may be seen as an aspect of historical research where it requires the researcher to examine how the building has been used over time. Here, there is a need to identify the context of the building, which may have changed over its lifetime. This may be done by comparison with other buildings of the period or by consideration of documentary evidence or other parallel artefacts.

Where these parallel sources do not exist, the researcher may rely more heavily on field notes, sketches or photographs made on the site. These may be compared with other sites investigated by the researcher.

In this form of investigation, the literature review and desk study may blend together or the literature review may identify sites and sources that need to be visited in person. Clear notes are needed and there must be an extensive evaluation which has an element of triangulation.

Referencing

If sections of the work are not referenced, they are assumed to be the original work of the researcher. If that is not the case and the work, ideas, etc. have been obtained from someone else, presenting that work as the researcher's own is plagiarism – intellectual theft – and is treated very seriously indeed. However, virtually everyone makes mistakes and occasionally omits references accidentally. Plagiarism can be thought of as omitting references deliberately; it has been found to the extent of copying and submitting someone else's entire dissertation. In such instances, penalties can be severe.

Referencing is important because:

- It demonstrates that the student has done the necessary secondary research and has an understanding of the subject.
- It allows the reader to find the original sources used in providing evidence for the arguments raised.
- It demonstrates that the student is not passing off someone else's research as their own thoughts and is, in fact, building the research on existing knowledge.

There are several standard methods for referencing. The Harvard system is used widely, and the guidance here is based on this. However, you should be aware of the requirements of your home institution, which may differ from this.

Whichever system is used, it must be consistent.

Briefly, a citation must be given which links to the full reference for the original work in the reference list in each of the following cases:

- if the work of another is used directly – as a quote or a copy of a diagram, for example;
- if the work of another is used indirectly – ideas or findings have been summarised in your own words or a set of statistics has been plotted in a new format, for example.

Citing sources

A citation acknowledges the work of others in the text without interrupting the flow. It is simplest to think of this as a system of cross-referencing between the sources in the text which has been written and a list of the sources from which the information was taken. The citation indicates to readers that a statement is based on secondary data and links to the reference list in order to give the information needed to find the source of the data.

Common knowledge need not be cited, but a citation should be given when information from any kind of source is used, even if it has been rewritten by the student (paraphrased).

It is good practice to provide the page numbers or page ranges with the references to save the interested reader from having to hunt through the reference in question to verify any statement made.

Examples of how to cite a source

Referencing conventions, as given below, are not universally prescribed and will vary between published sources and between (and within) academic institutions. The citation examples given below are fictional and given for illustration only.

Use the format 'author (date:page)' where the author's name flows as part of the text (see Example 1); and use the format '(author, date:page)' where the author/date information does not have a natural place in the text (see Example 2).

EXAMPLE 1

Black (2011:11) states '*referencing accurately is very important*'.

EXAMPLE 2

It is argued that '*an appropriate system adopted early will save significant time*' (White, 2011:4).

Dots (known as ellipses) are used to indicate that original words from the quote have been cut. Square brackets indicate to the reader that the author of the dissertation has added words to the quotation. In the following example, the square brackets show that the author has added words to make the fragments flow as a sentence: It is the case that '*without accurate referencing* ... [any source] ... *no matter how relevant, cannot be accepted*'.

Where a long passage is quoted, the text may be indented, as shown in Example 3. However, plagiarism software may remove some or all formatting, and thus it is recommended (though check if there is advice from your university) that quotation marks are also included for clarity.

EXAMPLE 3

> '*A long passage is not necessarily a good way of communicating to the readership that the content of a document has been read and understood. Any quote should be included only if the author decides that it summarises the work in such a way that paraphrasing would be a less effective way of presenting the text. Example after example of indented quotation divided only by a few student words is generally an indication of poor academic practice and low understanding.*'
>
> (Grey, 2015:12)

If a page number or range of page numbers is used in the citation, these are presented as follows: for a single page – 6; for a range of pages – 6–12; for a series of individual pages – 6, 8, 12.

Assembling the elements of a reference

The purpose of referencing is to create a means for a reader to find the source of the text used. As noted above, sources may include textbooks, journals, websites, etc. The specific details to be recorded for these different types of sources vary, but these include the name of the author(s), the date of publication, the publisher, and the place of publication. The generic information in Table 5.3 covers the basic information that is needed for different types of source.

Table 5.3 A guide for writing references

Type of reference	Format
Book reference Single author	AUTHOR, I. (Year) *Title*. Edition. Place: Publisher. Page.
Book reference Two authors	AUTHOR, I. & AUTHOR, I. (Year) *Title*. Edition. Place: Publisher. Page.
Book reference Three authors	AUTHOR, I., AUTHOR, I. & AUTHOR, I. (Year) *Title*. Edition. Place: Publisher. Page.
Book reference More than three authors	AUTHOR, I. *et al.* (Year) *Title*. Edition. Place: Publisher. Page. Or listing all authors (useful where the lead author has published several documents with different co-authors): AUTHOR, I., AUTHOR, I., AUTHOR, I., AUTHOR, I., AUTHOR, I. & AUTHOR, I. (Year) *Title*. Edition. Place: Publisher. Page.
Edited book	EDITOR, I. (ed.) (Year) *Title*. Edition. Place: Publisher. Page.
Chapter in an edited book Single author	AUTHOR, I. (Year) Chapter. In Editor, I. (ed.) *Title*. Edition. Place: Publisher. Page range.
E-book reference Single author	AUTHOR, I. (Year) *Title*. Edition. [Online]. Place: Publisher. Page.
Journal reference Single author	AUTHOR, I. (Year) Article title. Journal *title*, **Vol** (Issue), page range.
E-Journal reference Single author	AUTHOR, I. (Year) Article title. [Online]. Journal *title*, **Vol** (Issue), page range
Corporate author	COMPANY NAME (Year) *Title*. Edition. Place: Publisher. Page.
Anonymous author	ANON. (Year) *Title*. Edition. Place: Publisher. Page. Or by title: OXFORD ENGLISH DICTIONARY (2015) *Title of Entry*. Place: Publisher.
Web page (general)	AUTHOR or EDITOR (Year) *Title*. [Online]. Date of upload or latest update. Available from: http://.... (Accessed DATE).
Online Corporate author	COMPANY NAME (Year) *Title*. [Online]. Place: Publisher. Available from: http://.... (Accessed DATE).
EN (European standards)	EN (Year) Number, *Title*. Place: Publisher. OR EN Number (Year) *Title*. Place: Publisher.
Conference paper	AUTHOR, I. (Year) *Title*. In EDITOR, I. and EDITOR, I. (eds). *Title* of proceedings from the conference, Location of the conference, Date of the conference. Available from: http://.... (Accessed DATE).
Several works by the same author in the same year	AUTHOR, I. (Year a) *Title*. Edition. Place: Publisher. Page. AUTHOR, I. (Year b) *Title*. Edition. Place: Publisher. Page. AUTHOR, I. (Year c) *Title*. Edition. Place: Publisher. Page.

The details in Table 5.3 represent the information most commonly required to produce references so that the reader can follow up on the citations in the text. These should be adapted as necessary and expanded upon where additional fields are required in order to clearly identify the source.

Where the author(s) of a newspaper article is not identified, use the name of the paper as the author – for example, *The Times*.

If the author is not apparent on non-corporate websites, look for information on the copyright holder or look for contact information. If a date is not obvious, look for the date of the last update or the copyright date range.

If an organisation uses a standard abbreviation in its name (e.g. HSE for the Health and Safety Executive) then it is acceptable to use this in the reference list. The full name should be given the first time it is cited; for example, '... the item featured in the 2015 documentary by the BBC (British Broadcasting Corporation) has shown that ... ' Here, a citation in the form (BBC, 2015) can be added at the end of the relevant extract if there is a desire to be absolutely sure that the reference can be linked to the text. In the reference list, the abbreviation should be listed first (followed by the unabbreviated version where appropriate) to prevent the reader having to trawl through the reference list hunting for the document in question.

In some cases, certain elements of the reference will be unavailable. Where publisher details cannot be located, use the following conventions: if the publisher cannot be identified, use (s.n.) to indicate name unknown (from the Latin *sine nominee*); it is likely in such cases that the place of publication will also be unknown, and the student will therefore need to use (s.l.), meaning place unknown (from the Latin *sine loco*). Since (s.l.): (s.n.) is not helpful to someone trying to locate the reference, try to identify the publisher if at all possible.

A final point to note is that the citation used in the text must match exactly with the reference given in the list.

A little time spent understanding the basics can avoid common errors and will significantly improve the final quality of the work (see examples).

Review example 1

The process of good referencing is often seen as a process which is simple and yet many students seem to struggle to understand the formatting. Grey & Brown (2015:1) consider that this is understandable given the reliance on word processing. There is indication that slight changes in format can cause confusion (Tan, 2015:1), and several authors (Black, 2015; Green *et al.*, 2015; Blake, 2015b; Rojo, 2015) have suggested means to rectify these issues.

Critique

Good practice: several authors listed correctly and correlate with list given. Reads reasonably well.

Review example 2

'Clear referencing is the primary indicator of excellent research' (Shirosaki, 2015).

Critique

Good practice: Quote is clearly shown by use of speech marks and italics.

Poor practice: The page number for the quotation is not included in the citation. Also, Shirosaki is listed twice in the reference list with both publications in 2015. It is therefore not clear which reference this refers to; it should be (Shirosaki, 2015a) or (Shirosaki, 2015b).

Review example 3

In his book *The Importance of Making Sure that You Understand and Clearly Cite References: An Encyclopaedic Guide for Beginners*, Shiramine says *'Referring to a detailed textbook of references can be an invaluable aid to the new researcher. A book such as this needs to be read from cover to cover in order to fully understand the subtleties which are apparent. The demonstration of ability to collate and manipulate text using word processing or other electronic means.'*

Critique

Good practice: Quote is clearly shown by use of speech marks and italics.

Poor practice: No date or page number is given. It is not really necessary to repeat the book title as this can be found in the reference list. This text could be paraphrased: for example, Shiramine (2015:1) produced a reference guide which outlined good practice in collating references electronically.

Review example 4

Contrary to accepted thinking and the greater majority of published text, one source stated that this was not necessary in all cases: www.Blanco_ Blauwaert.com/Kuromine/bl..../Moreno.html.

Critique

Good practice: Has some means of verifying the source

Poor practice: Long URLs look untidy and take up valuable word count. Not in Harvard format: no author, date or page number. The reader wished to check the statement but could not locate the web page using the URL provided. Author, date and title would have increased the likelihood of locating the resource using a search tool.

Presentation of the reference list

The reference list needs to be accurate and consistent in style. It should be presented at the end of the document with sources arranged alphabetically by name of author (individual or corporate). Where the same author(s) have produced a number of references, these are ordered oldest to most recent or most recent to oldest (again, whichever option is followed, be consistent). Where the same author(s) have produced more than one work in the same year, these should be distinguished from each other using letters (a, b, c, etc.) included after the year in the reference list and in the citations in the text. Where a series of references have the same author and various co-authors, these should be listed alphabetically, initially by the first author, then by the second (and subsequent) authors and finally by date.

Whatever style is used for the different types of source, it should be consistent and comply with the requirements of the home institution. For example, if a decision is made to put book titles in bold or italic then care must be taken to ensure that all book titles are in the same format. If author surnames are in capitals then make sure this is carried on throughout. Similarly, while Table 5.3 recommends the use of initials for authors' forenames, these may be written out in full; but this should always be done consistently. If the full names of all authors cannot be found then for the sake of consistency, it is advised to use their initials only.

An example reference list for the work presented in review examples 1 to 4 is shown in Table 5.4 with a critique.

Table 5.4 Example references with suggestions to improve practice

Reference	Problem
BLAKE, S. (2015, unknown). Excellent references. Retrieved August 25, 2015, from References 4 U: www.references4U.co.uk	Years should read (2015a) and (2015b) as both references are by the same author in the same year. There is no need for 'unknown' in the first reference.
BLAKE, S. (2015). About references. Available from: www.AcademicSkillz.uk (Accessed 25 August, 2015).	The URLs should be for web pages referred to rather than home pages. Order of fields and terminology (e.g. 'retrieved'/'accessed') is not consistent between references.
BROWN, T. (2015, 17 August) Guidance. From Peacock Academic Services. Available from: www.PeacockAcademicSRVS.co.uk (Accessed 18 August, 2015).	The day/month ('17 August') can be put after the year if deemed important (the date of last update, for example).
GELBER, Y. (1015). *Timescales for research*. Journal of Timely Referencing. 4 (1), 8.	The date is clearly incorrect. Italics are used for journal article rather than for journal title – the style should be consistent with other journal articles in the list.
HUANG, Y., SILVER, V. S., KUROKAWA, N. I. & WEISSMANN, S. W. (2008). Issues Affecting the Use of Harvard by Students. *Referencing Weekly*. 2 (1), 1–7.	Italics are not used for the journal title – keep the style consistent.

Table 5.4 (Continued)

Reference	Problem
ROS, A. 2015. Writing good references. *Referencing Weekly.* 2 (1), 1–7.	These are not listed alphabetically by authors' surnames.
RAY, D. 2014. Referencing well. *Referencing Weekly.* 3 (1), 10–17.	The years are not in brackets, which is inconsistent with other references in the list.
Kuromatsu (2015)	This is not a complete reference – there is no title or publisher. It would not be possible to identify the source from just this information. The author's surname is not given in capitals as in the rest of list – be consistent.
Paul Plum, 2015, Plagiarism prevention through adequate advice. First edition. White and Orange.	There are several formatting inconsistencies here. The reference starts with the author's first name, followed by the surname. The author's surname is not given in capitals. The year is not in brackets. The book title is not in italics. There is no location for the publisher. It is only appropriate to include the edition if this is the second or later edition of the text.
Referencing 101 – A. GREY and E. BROWN	This reference is listed by title rather than author. There is no date and no publisher details.
ROJO, R. (2015) Naming and shaming – poor referencing practice. *Citation News*. 1(1), 3.	Different font and bold text is used.
Scarlett, 2015; Shirakawa, 2015; Redmon, 2015; Kurosaki, 2015; Schwartzenegger, 2015	This is a list of several authors, but none of them have titles or other information. This should be five separate references using capitals for author name and giving the full reference details for each source.
Source: data.aoyama.co.uk/ mustard/ akamatsu/rese…/referencing	This contains no author, no date and no access date, and the full URL is not given. This will be listed alphabetically under 'S' for source.
VERMELL, A. (2015b). <u>What's the game with referencing?</u> Referencing Weekly. 3 (1), 3.	Inconsistencies here include use of underlining for the article title and no italics for the journal title. There is no other reference to VERMELL, A. (2015) in the list so it is unclear why 'b' was added in this case.
{Problem: missing reference}	The following URL was given in the text: www.Blanco_Blauwaert.com/Kuromine/Moreno.html The reference should be given here in full: Blanco & Blauwert Consultants (2015) How to cite sources. [Online]. Available at: www.Blanco_Blauwaert.com/Kuromine/Moreno.html (accessed 18 August, 2015). A citation is also required in the text alongside or in place of the URL: (Blanco & Blauwert Consultants, 2015).

Researcher question
The reference list

Do you feel able to locate the information needed to create a reference and use this appropriately in your work?

Yes. I feel confident in my ability and will now make a decision on how to store the references I locate so that I am able to assemble my reference list based on the citations I use.

No. I am not confident in this area. I recognise that this is important and that I will need to review my practice regularly to ensure that it is appropriate. I will seek additional help and will undertake to review this section in future to support my work.

Supervisor guidance on secondary data collection and use

Guidance from Supervisor A

The investigation, analysis and composition of secondary data, more commonly referred to as a literature review, will require you to read and then gather, organise and analyse information.

Make sure that you set aside a good few hours at a time in order to undertake your collection of secondary data. This can be a highly absorbing process and requires you to utilise organisational skills from the off.

When undertaking to review literature in a constructive and meaningful manner, it should be borne in mind that there are two differing types of literature review. The first involves an initial 'trawl' of online and hard copy publications to assist in identifying and formulating the potential focus of a research investigation. Here you will need to read and read to gain improved understanding regarding potential research topics. This literature review commonly contributes to the 'introduction' section of the dissertation thesis. The second is a focused review of literature that addresses the specific aims and objectives of the research project. Only publications pertinent to the specific aims and objectives of the dissertation should be drawn upon. The identification of prominent key authors and researchers will assist in the initial undertaking of this task.

The initial review of literature should be carried out at an early stage of the dissertation research process to help you identify what has been written and researched regarding a potential topic. This literature review can serve to help the researcher formulate a research topic. This initial review of literature helps to inform the focus of the research and the development of research objectives. Student researchers can discover previous research investigations that they wish to repeat or build upon. An initial, broad review of published literature can also serve to raise specific questions or spark interest in a particular research focus. If your research focus and objectives are unclear then an initial literature review that involves reading, reading and more reading can certainly help to *inform* the development of a research idea.

A focused literature review should be undertaken once the research aims and objectives have been established. It is common for students to find that they are either overwhelmed with a wealth of information or, conversely, struggling to find any publications relevant to their research aims and objectives. In my experience, it is rare to find a valid dissertation topic where no published literature exists. It is a more common occurrence that where a research student cannot find relevant literature, this is due to the student not dedicating sufficient time, vigour and care to the investigation of published literature. A quick search of Google with zero returns does not constitute a sufficient search of published literature for secondary data relevant to the research aims and objectives. In this instance, you should not rush to inform your supervisor to say there is nothing published related to your research aims and objectives.

It is good practice to keep a record of your approach to reviewing literature for published secondary data. By documenting what you have done and what you plan to do, as well as which resources you have used and plan to use, you will be better able to review your approach and better informed when discussing and outlining your approach with your supervisor.

It is important to discuss your approach to your literature review with your supervisor. A supervisor may recommend specific journals or databases for you to search, or they may advise you to search for specific authors, experts or organisations. I regularly recommend students to spend time utilising the excellent ARCOM (Association of Researchers in Construction Management) website in order to search for built environment research publications (both conference and journal). A keyword search of the 'abstracts' section of the ARCOM website enables an initial search of a wide range of built environment–related resources. These resources include numerous journals as well as conference proceedings and PhD theses. The British Library's EThOS (E-thesis online service) database is another excellent resource for students wishing to read and investigate published PhD theses. EThOS permits students to search and instantly download published PhD theses. These are both excellent resources. I have also found Google Scholar to be of much value to a great number of students who are undertaking their review of literature. A whole wealth of journals and databases can be instantly searched online. This has certainly revolutionised the process of gathering published literature for use as secondary data within a dissertation. Once relevant publications have begun to be gathered, read and organised, the key references of these publications can also provide a fruitful further avenue of investigation.

Some research investigations are conducted entirely from the use of secondary data. Dissertations can also be conducted in this manner and these are sometimes referred to as 'desktop studies' or 'literature review dissertations'. Here, no primary data are collected but instead already published data provides the basis for the dissertation data. Dissertations that utilise such an approach may critically appraise existing case studies or compare and contrast the published data drawn from a number of research studies.

When a vast array of primary research data is available, this can be used in what is referred to as a 'big data' research project. Such projects utilise very large data sets to research occurrences and behaviours in order to develop and propose meaning, theories and algorithms. 'Big data' that exists in already published format may include sources such as death records, crime records or planning consents; indeed, any available large data set can provide the context for a possible secondary data desktop investigation.

My last bit of guidance regarding the review of literature concerns your focus – 'stay focused'. It is all too easy when reading and discovering new things to be diverted off topic for hours on end. Try to ensure that your review of literature and your collection, discussion and evaluation of published data remains closely related to the stated aims and objectives of your research.

Guidance from Supervisor B

Secondary data seems to fall into two camps. Either there is too much, or not enough. In both cases, the key aspect is not to panic and to identify issues early in the process so that you can take appropriate actions. If there is too much then consider what you are trying to do; are you communicating a development of ideas over time, or are you trying to put across the current state of the art? Consider how best to get the message from the literature over to your reader. While writing long tracts of text is the traditional way and generally highly appropriate, don't discount the use of tables or figures as well as or instead of this if it works.

In terms of sources, I thoroughly recommend looking at your university's online journal subscription pages. The use of keywords and dates can readily narrow down the search and produce a manageable volume of good-quality information. Using searches from the library may seem harder than using an online search engine, but it avoids such issues as suddenly encountering a paywall for an interesting article. In addition, I would personally avoid the overuse of resources such as Google Scholar as a sole source of information. While a good deal of the information accessible through there is of good quality, there is a lot which is out of date and this will show very clearly in your reference list. Just use any resource with care. Your evidence gathering here will be important later on when you discuss your results.

My top tips for searching would be to look at previous dissertations on similar topics, if available, and see what journals they are using for their sources. Make sure you know how to do a library search. If you need inter-library loans then order these in good time. If resources are held in another university library in physical rather than electronic format then you may even need to visit them 'in person', which has a time and money implication. Your supervisor may be able to help you by suggesting certain resources, but they probably will not give you a list of references and ask you to condense it down.

Guidance from Supervisor C

Being able to deliver a good literature review is one of the key skills expected of an undergraduate. I would expect a student to be able to collect resources, critically analyse them for content and be able to produce text in an appropriate academic format. The written work should offer a collation of the current state of the knowledge presented to me as a foundation from which the research will develop or mature. Good work can be produced entirely from the consideration and re-analysis of secondary resources; and this is overlooked in some fields whereas it is common in others. Where the secondary data are so vital, I would expect a significant reference list demonstrating understanding and linked by critical evaluation and insight into the findings.

Student reflections on approaches to secondary data collection

'Perhaps one of the overlooked facts is that the topic needs to have a healthy amount of resources which will support your proposal and provide a backbone to the study.'

'Choose a topic that you are able to get plenty of information on. Before you decide on a topic, briefly scan through some research databases for previously completed research in journals. I found this form of secondary information the most useful and, without a doubt, it produced the richest data to be convincingly used in my conclusions.'

'I wasn't sure what a literature review should look like. If I was doing this again, I'd make sure that I'd found out so I had a good idea what I was supposed to be doing.'

'Write up your references as you go along, as it will be surprising how many references you will use. Having a great piece of info that you can't include because you can't find the reference is very frustrating.'

'Literature-wise you need to keep reading the area, and if you find something interesting then make sure you further "probe" it even if you may not include it as it'll probably still be of interest.'

'Finding information in literature is easy, but making sure it's relevant isn't!'

'I was pleased with the library catalogue, particularly the journal database; there were many journals in my subject. The thing that worked the best was to spend around three hours one evening a week jotting down words that had the same meaning so it was easier to pick up research items on the Web and in the catalogue.'

'Try subscribing to any topic-relevant journals/magazines so regular emails are sent to you with brief news updates. It helps to make sure you do not miss anything that will help/make a difference to your research.'

'Make sure you don't spread your net too wide, otherwise you will drown in information and your dissertation won't focus; it will just end up being a mish mash!'

'Try and keep doing a little often rather than loads at once. I know that's easier said than done when you have other things going on, but if you get a couple of bits a week, you'll find it builds up quickly and you can get rid of the less relevant ones instead of just searching for anything to make up your reference list at the end.'

Summary

This chapter has briefly introduced secondary methods for data collection, which can be used as part of the methodology or as the sole source for data. The need to have a thorough understanding of the academic literature in order to underpin student work has been stressed. The academic conventions relating to citation and referencing have been overviewed, and suggestions have been made to help the novice researcher plan and execute their literature review.

Suggested further reading: approaches to data collection

DAWSON, C. (2009) *Introduction to Research Methods: A practical guide for anyone undertaking a research project*. 4th ed. Glasgow: Bell & Bain Ltd. ISBN 78-1-84528-367–4

FISHER, Colin *et al*. (2010) *Researching and Writing a Dissertation: A guidebook for business students*. 3rd ed. Harlow, Essex: Pearson Education Limited. ISBN 978-0-273-72343-1
Chapter 2: Writing a critical literature review, pp. 91–132.

GLATTHORN, A. A. & JOYNER, R. L. (2005) *Writing the Winning Thesis or Dissertation: A step-by-step guide*. Thousand Oaks, CA: Corwin Press. ISBN 0-7619-3961-X
Chapter 14: Mastering the academic style, pp. 135–54.

LOCKE, L. F., SILVERMAN, S. J. & SPIRDUSO, W. W. (2010) *Reading and Understanding Research*. 3rd ed. Los Angeles: Sage. ISBN 978-1-4129-7574-2
   Chapter 2: When to believe what you read, pp. 23–52.
   Chapter 3: How to select and read research reports, pp. 53–70.
RUDESTAM, K. E. & NEWTON, R. R. (2007) *Surviving Your Dissertation: A comprehensive guide to content and process*. Los Angeles: Sage. ISBN 978-1-4129-1679-0
   Chapter 4: Literature review and statement of the problem, pp. 61–85.
THE UNIVERSITY OF YORK (2012) *Reference with Confidence. The Harvard Style*. © 2012 Learning Enhancement Team. [Online]. Available at: https://www.york.ac.uk/integrity/downloads/15701_Harvard%20Style-webFINAL.pdf [last accessed 05/10/2015] 10pp

6 Research concepts

Introduction

This chapter outlines some key concepts that inform and shape dissertation practice. The chapter seeks to help dissertation researchers better understand a range of concepts that are necessary for the appropriate development and delivery of research projects. As such, this chapter concisely:

- introduces scientific and interpretive research perspectives and the meaning of research findings associated with the perspectives;
- considers quantitative and qualitative data;
- discusses some common approaches to data collection and sampling;
- identifies some key concepts by which research practice is judged – bias, rigour, reliability, repeatability, subjectivity and objectivity, and validity;
- provides guidance from three supervisors;
- offers views from students;
- suggests further reading.

Considering methods of investigation

When determining the methods of investigation to be deployed within a dissertation research project, the initial development process must consider and give reference to a number of factors. Such factors are interrelated and include:

- the aims and objectives of the research;
- the broad 'tradition' within with the research project sits – scientific or interpretive perspective (i.e. does the research project intend to produce findings that will be considered 'factual statements' or 'rich descriptions'?)

A research project that aspires to deliver statements of fact locates the research within the scientific perspective. A research project that aspires to produce rich descriptions of a particular event, occurrence or social setting but does not claim that findings are universal is defined as being located within the interpretive perspective.

Scientific and interpretive perspectives can each utilise quantitative and qualitative data; it is the inferred 'meaning' of the research findings that serves to clearly distinguish the two broad perspectives.

The scientific perspective

This is concerned with the development and testing of facts and commonly relies on statistical analysis. A scientific research approach necessitates precise procedural requirements and may place strict controls upon environmental conditions. Control is a key feature of scientific research. Findings from research conducted under precise experimental conditions and reproduced in repeat experiments can be generalised from the research to a broader context. Such findings are generally referred to as 'facts'.

The interpretive perspective

This is concerned with finding out about something in its natural context. Research projects that are grounded in an interpretive perspective do not exert control upon the subject or environment of the research. Findings and data that are resultant from interpretive research investigations are typically described as being rich in description.

The findings and outcomes of interpretive research are not 'generalisable'. The findings are 'specific' to the research context and can be considered to assist in developing or constructing understanding and meaning.

Interpretive research draws predominantly upon qualitative data but can also use data that are quantitative in nature. When determining methods of data collection for interpretive research, the methods of investigation should be guided by the aims and objectives of the research as well as consideration of whether data are going to be analysed using solely qualitative or quantitative techniques or a combination of both.

Qualitative and quantitative approaches to data collection

Whilst research and its various underpinning world views can be described and segregated in numerous ways, the data that are collected, analysed, discussed and concluded upon within dissertation research projects fall into two general camps: qualitative and quantitative.

Put quite simply, quantitative data are measurable and are commonly numeric in form. Qualitative data are rich in meaning and description and might be drawn from such sources as observations, interviews, discussions, drawings, photographs and written text.

A combination of qualitative and quantitative approaches to data collection and analysis can be employed when researching a social setting or a phenomenon. This may be the case when developing theory or measuring the outcome of an intended change imposed upon the environment or phenomenon, such

Table 6.1 Features of qualitative and quantitative approaches to research

Aspect	Quantitative	Qualitative
Research perspective	Scientific	Interpretive
Research approaches	Experimental	Ethnographic; Phenomenological; Grounded theory
Researcher role	Neutral	Immersed/may be participant
Research environments	Controlled and monitored	Naturalistic
Research data	Measurable and numeric Obtained from testing	Descriptive Emergent from the field of study
Research outcomes/ findings	Facts	Rich insights into social worlds, practices and phenomena
Scope of findings	Generalisable	Specific to the research context

as might be the case with an 'action research' approach. When a study draws upon both quantitative and qualitative data approaches, quantitative data can be rigorously and statistically analysed and supported with rich description and meaning drawn from the qualitative data.

Table 6.1 provides a summary of key features of qualitative and quantitative approaches to research.

Researcher questions
Considering methods of investigation

Have you made an initial decision on the balance of qualitative and quantitative research that will be undertaken?

Yes. I have made my initial choices although I recognise that these may change as my work develops. I understand that some research approaches are aligned more with qualitative methods and some are more associated with quantitative methods. I appreciate that the nature of the data will affect the analysis methods open to me. I am ready to continue.

No. I have not yet made my initial choice. I will read further before making my decision.

Types of quantitative data

Quantitative data can be drawn from experimental investigations, observation, case studies, questionnaires, literature and interviews. Methods of capturing and recording quantitative data vary greatly and must be carefully aligned to the purpose and objectives of the research. The type of quantitative data to be collected will inform the way in which the data need to be described, analysed and presented. Quantitative data types include the following:

Nominal data or categorical data

Categories can be grouped and labelled but there is no underlying order among the categories; in other words, ordering of the categories is arbitrary. For example, rocks can be grouped into 'igneous', 'sedimentary' or 'metamorphic'; participants may be labelled as 'male' or 'female' or some may 'prefer not to say'; participants' employment status might be described as 'unemployed', 'part-time', 'full-time', or 'self-employed'. A further example is a number which is used instead of a name but which has no significance in terms of order (e.g. numbers on football shirts).

Ordinal data

There is a logical ordering to the categories, but the difference between categories cannot be measured. One example is data collected using the Likert scale: strongly disagree; disagree; neutral; agree; strongly agree. Further examples are evidenced in degree classifications, the adjudged condition of second-hand books, and finishing places in a race.

Interval data

This is numerical data where the intervals between each value are equally split, but there is no true zero (i.e. zero is allocated arbitrarily). Examples include temperature on the Celsius scale, cash flow on a graph and time periods on a Gantt chart.

Ratio data

The ratio scale has a non-arbitrary zero value; examples include mass, length, height and distance.

Parametric and non-parametric data

The nature of the data collected determines which statistical tests are appropriate. Parametric statistical tests assume that the data come from an underlying Gaussian (normal) distribution. This assumption is not held for non-parametric data.

Parametric data

These data are derived from an underlying population that conforms to a normal distribution (i.e. more frequent values around the mean and very low numbers of occurrences at either extreme). Parametric data may be:

- *continuous* – where the scale of measurement used could be divided into smaller and smaller units; for example, kilotonnes, tonnes, kilograms, grams, milligrams, etc. (Figure 6.1);

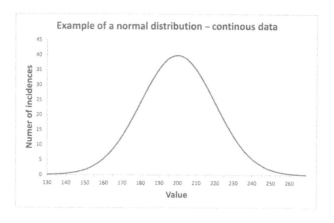

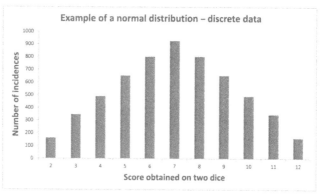

Figure 6.1 Examples of normal distribution for continuous and discrete data

- *discrete* – where the outcome is one thing or another – for example, the value obtained by two dice (Figure 6.1).

Non-parametric data

Non-parametric data do not make the assumptions of a distribution based on probability; for example, all ranked (ordinal) data are non-parametric (Figure 6.2).

As noted above, Likert scales, commonly used in surveys to collect attitudinal data, produce data that are ordinal. However, these are often treated as interval measurements for the purpose of analysis because it is assumed that the 'distance' between the points on the scales is equal (Figure 6.3). But this is not the case. Figure 6.4 illustrates a typical distribution for a set of responses. People are more likely to report a neutral response than to express a firm like or dislike. So even though a numerical value is obtained from the Likert scale, this should not be treated as interval data. This 'real' distribution cannot be known by the researcher for any group being studied.

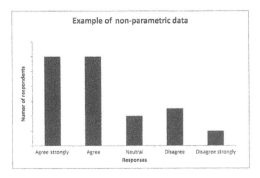

Figure 6.2 Example of a non-parametric data set

Strongly agree	Agree	Neutral	Disagree	Strongly disagree
1	2	3	4	5

Figure 6.3 Likert scale with the assumption of equal distribution along scale

Strongly agree	Agree	Neutral	Disagree	Strongly disagree
1	2	3	4	5

Figure 6.4 Likert scale with an unequal distribution along scale

Types of qualitative data

Qualitative data are commonly drawn from:

- observation of a social setting without direct interaction;
- participant observation – observation of a social setting with direct social engagement;
- interviews or questionnaires;
- written documents;
- drawings, photographs and other cultural artefacts.

Qualitative data can be recorded in a variety of ways. The methods of capturing and recording data very much depend upon the nature of the research and the resources available. It is good practice to ensure that such qualitative data are organised and are accessible at a future date for possible inspection by examiners.

Approaches to data collection

A number of approaches are available for collecting qualitative and quantitative data. These include:

- experimental approaches;
- case studies;

- ethnographic inquiry;
- interviews;
- questionnaires.

Experimental approaches

In simple terms, experimental approaches involve administering a treatment (e.g. showing a film to a group of students or applying pressure to an object) to an 'experimental group' and comparing the effects against a 'control group' (where no treatment is administered). If there is any difference between the two groups after the treatment, this can be attributed to the effects of the treatment. Experiments are commonly thought of as being laboratory based. This is not necessarily the case as quasi-experimental methods may be used in a diverse range of settings. An experiment may be carried out in order to:

- test a theory – if x is done, y happens;
- confirm an existing finding using the same or similar methods;
- build on previous work by testing an existing finding with different variables;
- contribute to the evidence base where previous studies in the literature have conflicting results.

Arguably, the greatest strength of experimental methods is that the procedures can be repeated. A well-designed and well-executed experiment should be able to produce results as prescribed in Table 6.2.

It is important that a researcher about to engage in experimental study in the field or the laboratory has a full understanding of exactly what will be required before commencing the work to allow for necessary changes in the proposed procedure to be made before the experiment commences. It is all too late to implement changes if it becomes apparent that the data collection methods are inadequate halfway through an experimental procedure. Table 6.3 outlines considerations that need to be given prior to conducting experimental research. (The practicalities of conducting an experiment are considered in Chapter 7.)

Types of experimental design

Experimental design refers to approaches for allocating participants to a control condition or an experimental condition. Table 6.4 briefly outlines some common

Table 6.2 Summary of key features of experimental research results

Rigorous	Trustworthy, following an appropriate strategy
Repeatable	Results can be reproduced with the same method
Reliable	Consistent and systematic
Valid	Represents what was intended
Objective	Data have been gathered and interpreted without subjectivity
Generalisable	Findings can be applied to a wider population

Table 6.3 Considerations to be made prior to conducting experimental research

Considerations	Comments	Refer to
How is this experiment linked to existing knowledge?	Literature must be consulted to inform the process	'The literature review' in Chapter 5
How is the sample to be derived from the underlying population?	Selection of research participants must be appropriate to the topic	'Sampling' in Chapter 6
What is the nature of the data to be collected?	For example, will the data collection involve testing inanimate objects or human subjects?	'Laboratory research' in Chapter 7
Are there any ethical considerations? Health and safety considerations?	Issues involving personal information or company-sensitive data should be considered, as should the well-being of researcher and participants	'Ethics' and 'Health and safety' in Chapter 7
What are the logistical considerations?	Time frames need to be in place for devising, gathering and analysing data	
What methods will be used?	What is to be studied and how? What methods are used in the literature?	'Laboratory research' in Chapter 7
How will the method be tested?	Need to ensure that questions are clear and unambiguous – a pilot test is vital	'Piloting and pre-testing' in Chapter 6
How will the data be analysed?	The number of responses, type of data, nature of the sample and other factors need to be considered	'Quantitative data analysis' in Chapter 8

Table 6.4 Common terminology associated with the experimental approach

Terms	Description
Independent variable	The variable which is controlled by the researcher (heat/light/sound) or which varies on its own (time)
Dependent variable	The variable being tested (strength/reaction/colour)
Specimen	An inanimate object
Participants	People in an experimental or quasi-experimental situation
Sample	A small part or quantity which is intended to be representative of the whole
Subject	A non-human animal

terminology associated with experimental approaches that need to be understood in order to start designing an experiment.

The experimental design adopted for a project is usually specific to the piece of work, and it is difficult to give guidance that will be relevant to every project in every case. The differing approaches to experimental design are shown in Table 6.5.

Planning an experiment

This section gives a brief overview of issues that may be considered in planning an experiment. Experimental approaches may have significant time demands associated with recording data. If monitoring and reporting times are low, it may be possible to increase the size of the sample. Likewise, if this process is time-consuming, the researcher may be forced to decrease sample size or change the scope of the study.

Supervisor knowledge, particularly where they have an interest in the field of research, should naturally be sought at an early stage. It is also worthwhile establishing early on whether there are specialist staff resources available to help you. Many universities have technical support staff who work with the equipment and methods that a student is intending to use. Consulting with them can save a vast amount of time. It may even be worth booking an appointment to talk through your proposal with these specialist staff before the research proposal is formalised as they will be aware of the availability of equipment and time requirements and may be able to identify potential problems before they impact on the project.

Planning the logistics of an experimental method over the given time and with the resources available is strongly advised. Preparation of a Gantt chart is recommended to clearly identify which areas of the experiment need to be carried out at which stages. Considerable time savings can be made by identifying procedures which can be carried out simultaneously. It cannot be stressed strongly enough that the time frame for conducting the experiment should be identified as early as possible, building in as much flexibility as can be managed.

The ethical issues and the potential health and safety implications of your proposed work should be considered and discussed with a specialist to ensure that suitable precautions are in place. Consult with your supervisor and/or technical staff with regards to safety procedures, such as the use of personal protective equipment (PPE), to ensure the execution of safe working practices in line with the home institution's policies.

Case studies

This is an approach to collecting data that involves the in-depth study of a particular 'case'. The focus of the case might be a project, an organisation, a department, a process or an activity. The case is investigated in its everyday context, and data can be gathered by use of various methods.

Table 6.5 Approaches to designing an experiment

Design	Sample allocation to test group	Test	Advantages	Disadvantages
Independent measures	The sample is divided randomly between the control condition (group 1) and the experimental condition (group 2) (e.g. members of the sample group allocated an odd number are in the experimental condition and those allocated an even number are in the control condition)	Group 1 is tested in condition 1 (control) and group 2 is tested in condition 2 (experimental)	Using a different sample in each condition means there are no order effects (where the same group takes part in more than one condition, results can be skewed due to fatigue or increased understanding of the experiment)	The sample needs to be twice the size of the sample in a repeated measures design to get the same amount of data The differences between experimental and control groups can distort results
Repeated measures design	The whole sample takes part in experimental and control conditions	Group 1 is tested in condition 1 (control) and group 2 is tested in condition 2 (experimental) Then the groups are swapped and the test re-run (with group 2 as control group)	Differences in the samples can be accounted for as both samples are tested in both conditions	Not always possible (e.g. where the measurement method destroys the sample) Order effects can distort results
Matched pairs	The sample is divided into pairs matched according to specified characteristics (e.g. age) One of each pair is randomly allocated to the experimental condition, with the other in the control condition	Group 1 is tested in condition 1 (control) and group 2 is tested in condition 2 (experimental)	Comparison of results between the two conditions is aided by the similarity of the two groups (based on carefully matching according to key characteristics)	Difficulty of matching pairs based on appropriate variable(s), which need identifying and measuring

The case study method enables the researcher to concentrate on a specific instance or situation and to identify and describe the various interactive processes that occur. These processes may remain hidden in broader large-scale studies.

Before conducting a case study, it is important to define the scope and focus of the case study if research aims and objectives are to be effectively achieved. The study must be methodically planned so that data are collected systematically and so that the relationships between variables can be studied.

Not all case studies are 'live' real-world investigations. Case studies can be constructed from secondary data sources as well as being based on primary data. Case studies that utilise primary data involve the study of phenomenon in their natural real-life context. Such case studies may deploy various data collection methods including observation, surveys, interviews and ethnographic inquiry.

Case studies that are based on secondary data utilise a literature review or desktop study approach and draw upon data provided by studies previously conducted. Secondary data are usually drawn from published literature sources, be these electronic or hard copy. Please note that a case study conducted via secondary data sources contained in published literature must always be reported as such.

The researcher defines the initial scope of the study in terms of what is to be investigated, how data are to be gathered and when. Generalisation of research findings is not usually sought; rather, case studies provide rich descriptions of the phenomenon that is being studied. (Some practicalities of conducting case studies in the field are outlined in Chapter 7.)

Ethnographic research

Research that utilises an ethnographic approach seeks to facilitate the production of in-depth descriptions of culture, practices and behaviours in everyday life settings. The ethnographic approach has its roots in anthropology, the study of the cultures of others.

Ethnographic inquiry requires the researcher to go into and observe or even become part of the social setting being studied. This social setting is commonly referred to as the 'field'. The researcher's engagement within the field varies in terms of the extent to which she or he takes the role of participant or observer.

- *Covert participant* – The researcher becomes an 'insider' in the social setting – someone who is accepted within the field as being a part of the group – without openly declaring their research interest. For someone conducting covert research, this approach may present significant ethical challenges.
- *Overt participant* – This describes someone who participates within the social setting but openly and overtly declares their researcher role.
- *Naturalistic observer* – As a naturalistic observer, the researcher does not

participate within the social setting. Here, the researcher only observes. Modern-day technology such as CCTV and webcams can facilitate naturalistic observation of geographically specific social settings, such as classrooms and community halls, without the researcher being present.

In undertaking ethnographic inquiry, the researcher must consider and contend with numerous issues which include:

- access to the social setting and their initial and ongoing acceptance within the field;
- their own role within the field (as an 'insider'/covert participant, overt participant or naturalistic observer);
- their impact upon the field (the 'Hawthorne effect');
- their means of documenting the research study in a timely, detailed, accurate and organised manner.

Ethnographic studies have been conducted and documented in various everyday social arenas and workplace settings. These include studies of construction sites (researched by Herbert Applebaum), police work (John Van Maanen) and railway engineers (Frederick Gamst). Such works provide detailed qualitative findings relating to the formal and informal aspects of social settings. Formal aspects can include rules, policies and prescribed procedures, whereas informal aspects can be described as the social interactions and interrelations of the people within the social setting.

Within construction industry settings, the field of study, when described in a formal manner, is made up of individuals with specific work-related roles and responsibilities. The roles and procedures of the workplace organisation can provide a useful and practical framework when attempting to document the structure, characteristics and interactions of this social setting.

Research data emerging from ethnographic inquiry can be qualitative or quantitative in nature. Researchers may document the number, frequency or structure of interactions, activities and occurrences within a social setting for the purpose of quantifying and measuring the social field and events therein. Commonly, ethnographic inquiry serves to facilitate the production of qualitative data. The researcher is the vehicle for the production of rich descriptive insights or emergent theory. Through the documenting and analysis of formal and informal actions, behaviours and conversations, the researcher is able to present a structured insight into the lifeworld of the social setting.

The researcher may intentionally move beyond the production of richly descriptive insights of the social setting to propose emergent theory or policy or practice changes relevant to the setting.

Within the context of an 'action research' approach to ethnographic inquiry, the researcher may choose to intentionally instigate changes in the formal or informal structures and practices of the social setting in order to monitor and document any resultant changes and outcomes. The action research approach of intentionally bringing about change and monitoring the out-comes can be described as experimental in nature and requires much ethical consideration.

Interviews

An interview is a verbal interaction between two (or more) people. The flow of information is from the interviewee to the interviewer. An interview is significantly different from a conversation where information flows in two directions. The interviewer will prompt the responses of the interviewee to a lesser or greater extent in order to meet the objectives of the data collection. When they are carried out well, interviews can provide extremely rich and detailed information.

Interviews are traditionally conducted face-to-face, but more recently there has been an increase in interviews by telephone or via video (e.g. using Skype).

Interviews are usually seen as offering more flexibility than questionnaires since the interviewer is able to prompt for areas of clarification or if further detail is needed. Data may be captured for later analysis in a number of ways (audio or video, transcripts of the spoken text, written notes from the interview). Table 6.6 provides an outline of differing types of interview.

Table 6.6 Types of interview

Type of interview	Characteristics	Advantages	Disadvantages
Fully structured	Fixed questions Fixed order Selection from a range of predetermined responses	Easy to analyse	More effort required initially in the writing of questions and responses Limited response options may cause frustration to participants Depth of data is poor
Semi-structured	May have some fixed questions May or may not have a fixed order Respondent answers are free-flowing	A compromise between ease of analysis and open responses that provide greater depth	Has some of the limitations of fully structured interviews in terms of the requirements of coding and limitation of scope
Unstructured	No fixed agenda	Control lies with the interviewee which can enhance feeling of collaboration between researcher and subject	An advanced technique which requires a significant commitment to data analysis Excellent depth and breadth of data Students are advised to discuss with their supervisors before selecting this technique

The interview sample

The time-consuming nature of carrying out and analysing interviews means that relatively small numbers of participants are involved. A highly targeted sampling approach is recommended.

For example, in a study of a particular field of work, compare interviews carried out with 5 selected practitioners in this field with 50 interviews with random members of the public. The former will result in less data; but because the sample is more targeted, it will offer more informed information. Of course, selection of the sample depends mainly on the research aims and objectives.

Table 6.7 outlines some considerations when planning to use interviews to collect data:

Questionnaires

The questionnaire is the method of data collection that is probably most familiar to students. Students are likely to have been recipients of various

Table 6.7 Considerations when using interviews to collect data

Considerations	Comments	Refer to
How is this research linked to previous studies or existing knowledge	Literature must be consulted to inform the process	'The Literature Review' in Chapter 5
What is the population to be sampled?	Respondents must be appropriate to research objectives	'Sampling' in Chapter 6
What is the nature of the data to be collected?	Data should be aligned to research objectives Need to know how many interviewees are needed	'Interviews' in Chapter 7
Are there any ethical or safety considerations?	This is a consideration for any research involving people but particularly if there are questions that involve the giving of personal information or highly sensitive company data	'Ethics' and 'Health and safety' in Chapter 7
What are the logistical considerations?	Consider the format of the interview, question setting, how it is to be recorded and how it is to be transcribed	'Interviews' in Chapter 7
How will the method be refined?	Need to ensure that the questions are clear and unambiguous – a pilot interview is needed	'Piloting and Pre-testing' in Chapter 7
How will the data be analysed?	Consider the type and amount of data, the nature of the data collected and the time needed to process and synthesise	'Qualitative data analysis' in Chapter 8

questionnaires. In many regards, this can attract students to the method before the full implications of carrying out this form of research have been considered.

Questionnaires can involve 'self-completion' or administration by the researcher. They may be filled in on paper, using an online survey tool, via an emailed document or over the telephone. Questionnaires are a relatively attractive method of data collection because of their low cost and the potential for inclusion of a larger sample than other methods such as interviews.

There are many aspects that have to be well thought through and suitably considered before questionnaires are developed and deployed. Table 6.8 presents an overview of considerations to be made when using questionnaires as a means to collect data.

Return rates for student questionnaires can be as low as a few per cent. It is essential to ensure that the design and delivery of the questionnaire is well thought through so as to encourage and maximise response. Table 6.9 provides a concise summary of suggestions for promoting a good response rate when using questionnaires. (These points are further expanded upon in Chapter 7.)

Table 6.8 Considerations when using questionnaires to collect data

Considerations	Comments	Refer to
How is this research linked to existing knowledge?	Literature must be consulted to inform the development process	'The literature review' in Chapter 5
How is the population to be sampled?	Respondents must be appropriate to the topic	'Sampling' in Chapter 6
What is the nature of the data to be collected?	The information should not be obtainable from other sources and must align to research outcomes Need to know how many respondants needed	'Questionnaires' in Chapter 7
Are there any ethical or safety considerations?	This is a consideration for any research involving people, especially if there are questions involving personal information or sensitive company data	'Ethics' and 'Health and safety' in Chapter 7
What are the logistical considerations?	Consider how questionnaires will be administered Consider the requirements of analysis and visual presentation of findings	'Questionnaires' in Chapter 7
How will the method be tested?	Need to ensure that the questions are clear and unambiguous – a pilot test is vital	'Piloting and pre-testing' in Chapter 7
How will the data be analysed?	The number of responses, type of data, nature of the sample population and other factors need to be considered	'Quantitative data analysis' Chapter 8

Table 6.9 Suggestions for promoting a good response rate in questionnaire data collection

Considerations	Suggestions
Relevance	The questionnaire should be of enough relevance to the potential respondent that they take suitable time to provide a response
Length	Questionnaires that are too short or too long may result in not enough data being collected Too short – higher rate of return but with less breadth of data Too long – the respondent may get bored before completion, leading to few returns or incomplete returns
Presentation	For self-completion options, the questionnaire needs to be laid out in a clear and attractive way
Sequence	Start with general questions and lead on to more specific and detailed questions
Objectivity	Questions should be written in a neutral voice

A follow-up or contingency plan should always be considered to increase return rates. Conversely, carefully consider the volume of questionnaire distribution. A highly successful response rate can leave a student researcher with the positive problem of struggling to cope with inputting the volume of data generated by a high rate of returns.

Researcher question
Approaches to data collection

Have you made an initial decision on the data acquisition type(s) you intend to use?

Yes. I have made my initial decision, although I recognise that this may change as my work develops. I have read the sections and can see how these are aligned to the purpose of my research. I also understand the strengths and weaknesses of each method.

No. I have not yet made my initial decision. I want to understand the methods further. After finishing this chapter, I will read the relevant sections of Chapter 7 on the practicalities of data collection before returning to this section.

Sampling

It is a relatively rare occurrence that any investigation is able to study all possible incidences over all times and in all places. It is therefore more likely that a sample will be derived for study. For investigations involving people, the

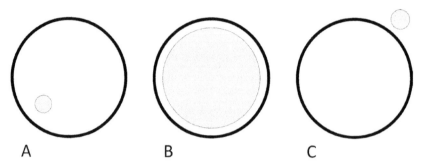

Figure 6.5 Relationship between sample and population

members of the sample are referred to as 'participants'; for laboratory investigations, individual items are used to derive 'specimens' for testing.

In some cases, samples are selected with the intent of making assumptions about
the behaviour of the whole population. In such cases, attempt is made to make
the sample representative of the population so that results are 'generalisable'.
Information such as population size and characteristics are used to define the best
strategy for deriving a sample. The larger and more representative the sample, the
greater confidence a researcher has in the generalisability of research findings.

In Figure 6.5, diagrams A, B and C illustrate examples of target populations
(bold circles) and the sample populations (shaded circles). The relative sizes
of the shaded circles can vary depending on the percentage of the population
studied. It can be seen that:

- In A, a large underlying population has a small sample taken from it. Care is
  needed to ensure that this is representative.
- In B, almost every individual within a population has been sampled, which
  is possible where there is a small population.
- In C, sampling has captured individuals who are not part of the target population. The sampling has failed as the sample does not represent the population.

Sampling methods

Sampling methods can be classified into:

- *Probability sampling* – There is a sampling frame (a list of members of the
  population) available and each individual has a known or 'equal' probability
  of being selected, and it is possible to select a sample that is representative
  of the population;
- *Non-probability sampling* – There is not necessarily a sampling frame and
  individuals do not have an equal probability of selection; a representative
  sample will not necessarily be chosen.

A good sampling strategy can be time-consuming and expensive. A poor sampling strategy will result in data that are unreliable at best and unusable in

Table 6.10 A selection of probability sampling methods

Source of possible bias	Means of mitigation
Literature review – selection of reading material	Try to present a balanced argument rather than a single-sided view Be thorough and objective
Study design	Clear definition of aims/objectives/hypothesis Select accepted and appropriate methods of data collection
Sampling	Use rigorous criteria in selection Aim for an appropriate sample size
Data collection tools	Standardisation of measurement methods Calibration of equipment
Data collection process	Standardisation of interactions with participants when collecting data Avoid giving 'clues' that may influence participant responses Take an ethical approach (do not pressure participants)
Quantitative data analysis	Take into account all variables Use robust data analysis techniques
Qualitative data analysis	Descriptions, categories and meanings should emerge from the research study and not be imposed by the researcher Ensure the presentation of qualitative research includes relevant supporting quotations and observations drawn directly from the primary data Researcher should be 'reflexive' and any personal biases acknowledged in the dissertation
Writing up and communication of results	Report all results, even those which are contradictory or unfavourable

the worst case scenario. A researcher therefore needs to consider the time and resources they have available when determining a sampling strategy.

Sampling for research that involves human participants is the focus here. Tables 6.10 and 6.11 outline, in turn, some probability and non-probability sampling methods. For sample creation within the context of experimental laboratory work, please refer to Chapter 7 and the 'Laboratory research' section therein.

Determining the size of a sample

For questionnaires and where the population is known, the size of the sample needed to give results that represent the population (within a specified level of certainty) can be calculated. This can be done manually or using readily available online calculators on statistics sites. If the sample size looks unfeasible at the design stage then consult a supervisor and determine the compromises that will be needed. Remember that an undergraduate thesis is significantly constrained by time and that the level of information collected will not be as extensive as for a full-scale PhD research project.

Table 6.11 A selection of non-probability sampling methods

Dissertation aspect	Demonstration of rigour
Hypothesis/aims and objectives	Ensure that these are clearly communicated and are addressed by the research.
Literature	Obtain, identify and analyse relevant existing knowledge and practice relating to your research. This should include authoritative sources and publications from a range of relevant sources.
Methodology	This must be fully expressed and detailed and should enable others to fully understand what has been done and how issues of validity, bias and reliability have been addressed.
Results	Results and findings are clearly linked to the hypothesis/aims and objectives. There is clear transparency in the processes of data collection and analysis. Results should be grounded in the research and presented in a clear concise manner.
Discussion	This should include consideration of the results achieved by the research and comparison of these results with those of research investigations reported by others. Include possible explanations for significant discrepancies and differences and suggest meaningful further research.
Conclusion	This should be accurately reported and include all outcomes, including those that were unexpected.
References	All sources are identified and included and presented in an entirely correctly manner.

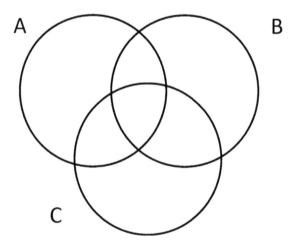

Figure 6.6 Areas of commonality and difference between three interviewees

For interviews, consider that the time needed to undertake the data collection is significant. A rule of thumb is to consider gaining three or more interviews from each subgroup of interest. The Venn diagram in Figure 6.6 illustrates this. The areas where all respondents concur and where their opinions differ from each other can be established.

Key concepts that inform dissertation research practice

A number of important concepts inform and shape research practice. These concepts include

- bias
- reliability
- repeatability
- rigour

- sampling
- subjectivity
- objectivity
- validity.

These concepts are concisely outlined below.

Bias

Bias is the presence of influences that can affect the study outcome. Bias can be present in any stage of a research project.

In scientific investigations, bias can occur due to a lack of systematic approach. It is essential that bias is removed through the application of standardised methods of data collection and strict control of research processes.

In interpretive investigations, the researcher is the analytical filter. Bias is mitigated by the researcher making every effort not to allow their own subjective views to influence the collection, analysis and presentation of data. Table 6.12 outlines aspects of possible bias that may be present within dissertation research.

If the researcher allows their own views to infiltrate the collection, analysis and presentation of qualitative data without acknowledgement or understanding of the effects on research outcomes, then the result can be thought of more as a subjective journalistic account than a research investigation. The researcher should try to mitigate the effects of bias on the research process and outcomes by acknowledging the existence of bias and documenting their involvement with the social environment or phenomenon of study. This type of personal account, termed a 'reflexive' account, serves to lay bare the researcher's own experiences, thoughts, and relationships with those being researched.

Reliability

Reliability refers to the consistency and dependability of tests. Only if a method is reliable can the results be repeated.

External reliability concerns the consistency of measures over time; they should produce the same data if used with the same people at different points in time. This is also known as test-retest reliability. Internal reliability refers to the consistency within the measures used. In a questionnaire, if questions addressing similar phenomena produce results that are inconsistent with one another, this means that internal reliability is poor.

Table 6.12 Sources of possible bias within dissertation research

Method	Description	Use	Advantages	Disadvantages
Random sampling	Each member of the population has an equal chance of selection	Large populations	Sample represents the target population and sampling bias is low	Very difficult to do – a list of every individual in the population is needed before sampling starts
Stratified random sampling	Population to be sampled is divided into subcategories and random sampling is done with respect to each of these groupings	Populations where proportions of subcategories are key to the sample being representative	The sample is highly representative of the target population	Time-consuming and difficult – a list of every individual in the population and their characteristics according to the subcategory is needed before sampling starts
Systematic sampling	Individuals are selected in a logical way (e.g. every third person on a list of participants)	Large, known populations where a set number of participants is required	Should provide a representative sample	Very difficult to do – a list of every individual in the population is needed before sampling starts

Repeatability

The concept of repeatability refers to the possibility of taking a method and duplicating it to establish if the same results are obtained. Repeatability is a theoretical consideration, whereas replication is actually undertaking the research on further occasions.

Both repeatability and replication serve to test whether assumptions made are justifiable and whether methods and findings have been reported rigorously.

Rigour

All research investigations start with a question and rigour relates to the way in which the answer is derived and presented.

Where a question is asked which uses a quantitative methodology, rigour relates to the testing of a hypothesis. Have the data been derived and analysed in such a way that the answer obtained is correct?

Rigour in qualitative research is probably easiest thought of in terms of the appropriateness of data collection techniques, the transparency and thoroughness in the carrying out of analysis and the trustworthiness of findings. Rigour in qualitative interpretive research is often promoted via the researcher acknowledging their own relationship with the social environment or phenomenon of study via a written 'reflexive' account of their experiences, thoughts, and relationships with the researched.

Rigour should be evidenced throughout the whole research dissertation process. Table 6.13 outlines dimensions of rigour in the presentation of dissertation research. This table can prove a useful tool to researchers when giving thought and due consideration to rigour.

Objectivity

Objectivity centres on the idea that there is a truth or reality which exists regardless of whether it is observed or not. The researcher must act in such a way as to discover this reality without contaminating it.

Objectivity is drawn from a positivist world stance, which is the default philosophy in the natural sciences where quantitative research methods dominate. The existence of absolute objectivity is disputed, however, and not fully or entirely supported outside of the 'scientific' world view. Objectivity is to be aspired to, but some subjectivity may be present; for example in the selection of samples, the types of measurement undertaken and the forms of analysis carried out.

Subjectivity

Within the 'interpretive' tradition adopted by some researchers, subjectivity is an accepted world view. As such, researchers who adopt a subjective interpretive perspective recognise that people are much more than observable objects and see the social world as being subjectively experienced, understood and created. It is held that the relationship between researcher and researched can never be objective because 'truth' is subjectively constructed by interactions between individuals. In this tradition, internal bias and preconceptions are acknowledged and documented.

Validity

Validity refers to how well a study examines the phenomena at which it was aimed.

In quantitative studies, validity is linked to the robustness of the research methods and the resulting findings. Researchers in the scientific tradition are concerned with the 'internal validity' of experimental research. This is not linked to research findings but, instead, concerns the design of an experimental investigation. When an experiment is internally valid, it means that changes in measured variables occur as a direct result of the intended changes made to the independent variable. Researchers also consider the 'external validity' of findings in experiments. This refers to the extent to which findings can be generalised to other research samples and settings.

In qualitative studies, the term 'credibility' often replaces the term validity. Qualitative investigations seek to present descriptions, insights and claims that can be considered to be credible, plausible and acceptable.

Table 6.13 Dimensions of rigour in the presentation of dissertation research

Method	Description	Use	Advantages	Disadvantages
Volunteer	Individuals are asked to volunteer (self-selecting)	Use when there is difficulty identifying members of a population	Convenient and ethical (informed consent)	Unrepresentative as the population is biased to those with an interest
Opportunity sampling/ convenience sampling	Individuals are available at a particular place or time	Use when there are time constraints	Quick, easy, convenient and economical Commonly used	Sample is unrepresentative of the underlying population
Quota methods	The researcher predetermines the subgroups and number of participants in each category and individuals are selected using an opportunity sampling approach	Use when access to a wide population, including subgroups, is possible and where a study aims to investigate a trait or a characteristic of a certain subgroup	The sample taken has the same proportions of individuals in each subgroup as the entire population	As only the selected traits of the population are taken into account in forming the subgroup categories, other traits in the sample may be overrepresented
Purposive sampling	Study is focused on a particular group	Use where there is a focus on certain characteristics or behaviours of a population	Provides highly focused results	Researcher judgement forms the sample and therefore this is subject to researcher bias
Expert sampling	Study is focused on experts in a particular field	Use when experts (those with a relatively high level of skill or knowledge) are needed for the study	Expert opinions are respected, adding credibility to the research	Requires the identification of 'experts' by the researcher and is therefore open to bias
Snowball sampling	Study is on a particular group with a small and unknown population	Use where there are difficulties accessing the sample but where an individual belonging to the group being studied can be identified and may be able to recommend others within the subgroup	Possible to include people that would otherwise not be identifiable as there are no lists or obvious sources of information on the population	Participants in the research have an impact on sample selection outside the control of the researcher May be time-consuming
Diversity sampling	Study aims to include maximum diversity of views/experiences within the sample	Use when differences between two or more subgroups are important for the aims of the study	Contributes to discovering and understanding variation	The final sample size in such studies may not be determined, although a minimum size may be identified

Supervisor guidance on research concepts

Guidance from Supervisor A

Reading and thinking about research concepts can be a time-consuming discipline in itself. Students can become overwhelmed with the notion that they do not fully understand all research concepts and their implications.

Grappling with the many research concepts and then identifying, understanding, discussing and applying those relevant to the dissertation can be a very challenging task. I advise students to take an approach that involves: reading selective chapters of key texts; and ensuring that they attend research methods lectures and then participating in a discussion session with their supervisor. I try to run this session as a group discussion with a small 'collective' of dissertation students that I am supervising and myself. The students are invariably researching various topics with differing aims and objectives. Regardless of each student's topic of research, I expect the students to engage in discussion regarding scientific and interpretive approaches to research as well as the notions of 'validity', 'reliability' and 'rigour'.

It is common for students to 'discover' for the first time the concepts of 'ontology' and 'epistemology'. In my experience, students can also believe they are expected to hold a clear resolute position as to what their own views are regarding ontology (what is 'reality'?) and epistemology (what is knowledge?).

Students can sometimes be overwhelmed at discovering a whole new world of philosophical discussion. My guidance to students is to be pragmatic – you should *identify and concisely summarise* the tradition within which your research sits. You also need to ensure that the assembled methods of data collection and analysis align and do not conflict with the stated tradition of the research.

When it comes to research that is undertaken within the context of the built environment, research can be located broadly within two traditions, these being the scientific perspective of logical positivism and the interpretive perspective.

Guidance from Supervisor B

I think that the challenges presented by the concepts related to laboratory work or fieldwork are mostly readily overcome. Students have most control when carrying out a laboratory experiment – the timings and logistics can be worked out with a high degree of certainty. The nature of the knowledge and how it is measured relates to the scientific paradigm, and a main consideration is the relevance of the means of measuring and its accuracy. Data are usually measured at interval or ratio levels and are, therefore, amenable generally to parametric statistical analysis techniques. Most analysis procedures are standardised and there is usually plenty of literature that enables findings or theories to be identified or theories tested, or existing work to be replicated, built upon or challenged. Identifying the samples and planning in adequate time (and contingency time to replicate or repeat any part of the procedure) will help. I encourage my students to speak to technical specialists to get an appreciation of what will be needed and when their work can be scheduled in. Laboratory time and resources can often be limited. Get a timetable for activities and stick to it. Raise issues early and deal with problems proactively. Check the logistics of your plans which must include consideration of the health and safety of yourself and others.

Take the same care if you are undertaking fieldwork; preparation is again the key. Make sure you know what data you need to collect (and why) how this will be collected. Planning is very much key. If any of your work involves people then you will need to consider ethical considerations, particularly if working in an organisation.

In all cases, make sure you know how you will analyse the data before you collect it!

Guidance from Supervisor C

It is important to prepare yourself for your dissertation research project by ensuring that you have a good grasp and understanding of key research concepts. If your research is to be developed, delivered, analysed and presented in an appropriate, acceptable and meaningful manner then it is essential that the project is built on a solid foundation of knowledge regarding research concepts.

Without knowledge and consideration of concerns such as bias, rigour, reliability, sampling and validity, it is virtually impossible to construct and undertake a credible, meaningful research project. These concerns or concepts have to be addressed from the outset of the research design, at the project proposal stage. They cannot be retrospectively attended to halfway through project delivery or 'retrofitted' to the project after data collection and analysis.

I suggest that dissertation researchers consider research concepts as applied to their own potential research investigation and discuss the topic with their supervisor. This should be done when beginning to design, develop and refine an approach to a research investigation at the project proposal stage.

Holding a clear understanding of research concepts and how they impact upon your dissertation research at the outset of the development of your dissertation proposal will help you ensure that your research design is appropriate, robust and well thought through.

Student reflections on research concepts

'All of my research was sequentially linked so that the feedback and outcomes from my previous researches helped shape the next so that comprehensive and enriched data were obtained. As I was distanced from my dissertation topic, I found it hard to find interview and case study contacts; however, by using a questionnaire, I gathered a vast array of further research possibilities just by simply asking participants whether they would be willing to be interviewed and if they had any relevant case study examples.'

'I remember looking at some past MSc dissertations and finding one that was just full of all these long words I had no idea about. I went to my supervisor and asked if that was the way I was supposed to write, and he told me that I would be better off with a pragmatic approach. Once I could get my head around what I was supposed to be doing, the rest of it made sense. My work got a good mark even without all the jargon.'

'I found the case study approach particularly useful as it allowed me to get an in-depth understanding of how one particular company operated site waste management.'

'Using a mixture of primary data collection media, such as questionnaires, interviews and case studies allows triangulation. You should contact people early to ensure you get responses.'

'Read up on technique when formulating your questionnaires and interviews. Don't try to do too much.'

'When you are designing your questionnaire, make sure the questions you are asking are going to be conclusive and think about how you will analyse the data when you are thinking about what data you might collect. If you're not good at maths, etc., make sure your data conveys a qualitative tone and vice versa. There is nothing worse than having a question which you can't analyse effectively.'

'Don't do too much! Don't be too ambitious with your data collection; you have word count and resource constraints, and by doing too much, you end up having to just omit some data that you have put time and effort into collecting.'

Summary

This chapter has introduced scientific and interpretive research perspectives and has outlined the meaning of research findings associated with these perspectives. This chapter has also considered quantitative and qualitative data and has identified some common methods of data collection. Finally the chapter has identified and presented key concepts that inform research practice.

Suggested further reading

BELL, J. (2010) *Doing Your Research Project: A guide for first-time researchers in education, health and social science.* 5th ed. Maidenhead: McGraw-Hill Open University Press. ISBN 978-033523582-7
Chapter 10: Diaries, logs and critical incidents, pp. 177–90.

CLEGG. F. (1983) *Simple Statistics: A course book for the social sciences.* Cambridge: Cambridge University Press. ISBN 978-0521288026

CRESWELL, J. W. (2012) *Educational Research: Planning, conducting, and evaluating quantitative and qualitative research.* 4th ed. Boston MA: Pearson. ISBN 978-0-13-261394-1
Chapter 7: Collecting qualitative data, pp. 204–35.

DAWSON, C. (2009) *Introduction to Research Methods: A practical guide for anyone undertaking a research project.* 4th ed. Glasgow: Bell & Bain Ltd. ISBN 978-1-84528-367-4
Chapter 5: How to choose your participants, pp. 48–56.

DENSCOMBE, M. (2010) *The Good Research Guide for Small-Scale Social Research Projects.* 4th ed. Maidenhead: Open University Press/McGraw Hill. ISBN 978-0-335-24138-5

FIELD, A. & HOLE, G. (2003) *How to Design and Report Experiments.* London: Sage. ISBN 978-0-7619-7383-6
Chapter 3: Experimental Designs, pp. 54–106.

FISHER, C. et al. (2007) *Researching and Writing a Dissertation: A guidebook for business students.* 2nd ed. Harlow, Essex: Pearson Education Limited. ISBN 0-273-71007-3

GILL, J., JOHNSON, P. & CLARK, M. (2010) *Research Methods for Managers.* 4th ed. Los Angeles; London: SAGE. ISBN 978-1-84787-094-0
Chapter 8: Philosophical disputes and management research, pp. 187–213.
Chapter 9: Evaluating management research, pp. 214–39.

GROAT, L. N. & WANG, D. (2013) *Architectural Research Methods.* 2nd ed. Chichester: Wiley. ISBN: 978-0-470-90855-6
Chapter 6: Historical research, pp. 172–214.
Chapter 12: Case studies and combined strategies, pp. 415–52.

HERR, K. G. & ANDERSON, G. L. (2014) *The Action Research Dissertation: A guide for students and faculty.* 2nd ed. Thousand Oaks, CA: Sage Publications Ltd. ISBN 9781483333106

HOLT, G. D. (1998) *Guide to Successful Dissertation Study for Students of the Built Environment.* 2nd ed. Wolverhampton: Built Environment Research Unit, University of Wolverhampton. ISBN 1-902010-01-9

KNIGHT, A. & RUDDOCK, L. (2008) *Advanced Research Methods in the Built Environment.* Chichester: J. Wiley & Sons Ltd. ISBN 978-1-4051-6110-7

MARDER, M. P. (2011) *Research Methods for Science.* Cambridge: Cambridge University Press. ISBN 978-0-521-14584-8
Chapter 2: Overview of experimental analysis and design, pp. 18–50.

MELOY, J. M. (2002) *Writing the Qualitative Dissertation: Understanding by doing.* Mahway, NJ: Lawrence Erlbaum Associates.

NAOUM, S. G. (2007) *Dissertation Research and Writing for Construction Students.* 2nd ed. Oxford: Butterworth-Heinemann. ISBN 0-7506-8264-7

OPPENHEIM, A. N. (1992) *Questionnaire Design, Interviewing and Attitude Measurement.* London: Continuum.

PIANTANIDA, M. & GARMAN, N. B. (1999) *The Qualitative Dissertation: A guide for students and faculty.* Thousand Oaks, CA: Corwin Press. ISBN 978-0803966895

Chapter 7: Living with the study: generating knowledge through portrayals, pp. 129–55.

Chapter 8: Living with the study: getting to portrayals, pp. 156–85.

ROBSON, C. (2011) *Real World Research*. 3rd ed. Hoboken, NJ; Chichester: Wiley. ISBN 978-1-405-18240-9

Chapter 10: Surveys and questionnaires, pp. 236–77.

Chapter 11: Interviews and focus groups, pp. 278–301.

SWETNAM, D. (2001) *Writing Your Dissertation: How to plan, prepare and present successful work*. 3rd ed. Oxford: How To Books Ltd. ISBN 1-85703-662-X

Chapter 4: Techniques, pp. 51–63.

7 The practicalities of primary data collection

Introduction

This chapter includes:

- the practical implementation of data collection;
- the ethical and health and safety controls which should be considered;
- factors which affect the selection and development of primary data collection methods;
- suggestions to improve student processes in development of primary data collection methods;
- comments from virtual supervisors on data collection;
- the views of students on this part of the dissertation;
- suggested reading.

The challenges of data collection

Primary data collection can be one of the most rewarding parts of undertaking research. In completing this process, a student can learn how real-world experiences inform current practice. The practice of undertaking any systematic investigation is not without risks, but these can be minimised by recognising them and acting to reduce these early in the process.

The principle areas of concern that should be addressed include those which could impact upon the researcher or their participants which require careful consideration of research ethics and health and safety. Although addressing ethical issues is sometimes seen by novice researchers as an additional burden, it is important that the student recognises the benefits which can be gained by following best practice. Considering possible sensitivities and minimising them will almost always improve the research process. In addressing health and safety issues and how these may be reduced, the researcher is encouraged to give consideration to their methods and logistics, which again will help both with the planning of the work and with the quality of the data gathered.

From a student point of view, the principle concern is often the failure of a particular collection method to gather the volume of appropriate data. Students are encouraged to design their methods carefully and with appropriate contingencies. While there are no guarantees that a well-designed and well-executed project will go smoothly, there is significant evidence that a poorly conceived and badly planned project will cause significant difficulties throughout.

Ethics

It is the researcher's responsibility to ensure that ethical issues have been addressed throughout the project. At all stages, there should be documentary evidence showing that due consideration has been given to potential ethical issues arising from carrying out the research project. Ethical consideration must be given in order to:

- protect the research participants;
- protect a researcher from litigation or unfounded allegations of malpractice;
- reassure the public that the research is ethical;
- meet requirements of insurance cover.

Clear planning and management of ethical issues serve to mitigate negative, harmful outcomes of the research process and improve the quality of the overall research strategy by forcing consideration of methods and practicalities in advance of research delivery.

All research should be done to comply with good practice. The Research Council UK produces guidelines on good research conduct (primarily at postgraduate level). Universities which are signatories to the concordat make a commitment to support research integrity by considering how work impacts on society (Universities UK, 2012). Key aspects are:

- participant selection;
- protecting research participants;
- obtaining informed consent;
- respecting confidentiality of participants;
- respecting the anonymity of participants;
- storing data securely;
- disposing of data in an appropriate way;
- ensuring the integrity and quality of the research;
- protecting the researcher.

A detailed ethical review is needed if the research undertaken involves direct contact with human participants (including social research). It is the researcher's responsibility to undertake this review and to share it with their supervisor, since the supervisor can only become aware of the researcher's intentions in this way. In addition, the student will need to be able to demonstrate compliance with legal requirements (e.g. data protection legislation).

Concepts relating to ethical practice are outlined in the following section.

Non-malfeasance and beneficence

A participant in research should be protected from harm (physical harm or emotional distress). The participant's interests are always more important than the research. All risks should be assessed before research commences and efforts made to reduce these to an acceptable level (ideally nil). If these cannot be predicted, the work should not go ahead. Procedures should be in place to deal with any potential hazards, and risks to the environment should be minimised.

Integrity

The research should demonstrate integrity by being soundly constructed and executed well. Supervisors may undertake verification of student data sets (examining data from the field or laboratory, listening to audio recordings of interviews, reviewing photographs). A clear audit trail of documents should exist.

Informed consent

Each potential participant must be adequately informed about the research, the methods, the possible risks and the anticipated outcomes.

Documentation given to a participant should be easy to understand. Participants should be given opportunity to read this and raise concerns or withdraw from the research. The form should be suitable for the type of data collection being carried out:

- For an interview this may be in letter form, given to the participant prior to meeting. A signed consent form from the participant should be securely stored by the researcher for the duration of the project.
- For a questionnaire, this may be in the form of a short paragraph at the start of the questionnaire.

Participants who can be identified by the researcher should always be given the mechanism by which they can withdraw their participation. While they should be able to withdraw at any point, practically removing data from a submitted dissertation is not possible; a reasonable cut-off date for withdrawal may be specified (e.g. three weeks before the hand-in date).

Good practice includes the following:

- Withdrawal may not be possible with an anonymous questionnaire. The participants need to be told that submitting the questionnaire is taken as providing informed consent.
- The information supplied to the participants would usually include the supervisor's details and the details of the home institution.
- Any incentive to participate should be declared and be appropriate.
- Fundamentally, any promises made to participants on a consent sheet must be feasible and put into practice.

- If techniques involve methods where participant consent cannot be gained, any information collected should be stored and handled as if it contained identifying information, and care should be taken to ensure participant anonymity during communication of findings.
- If the research involves minors, a student should consult with their supervisor as additional care is needed.

Confidentiality/anonymity

Research involving people is usually based on the participants remaining anonymous and data being treated as confidential. Any research which collects identifying data should conform to data protection legislation, and the data should not be made available to anyone not involved in the research without participants' consent. Data should be secured and, ideally, stored on university servers. Anonymised data should be used in the written dissertation. Naming a participant is sometimes necessary; for example, when their involvement is to be fully acknowledged as contributing to the research. In all cases where a respondent is to be identified, explicit consent must have been given to the researcher.

Since the researcher is generally in a position where they know the identity of their participants, it is important to gain informed consent should any audio or video recording be undertaken. This will require briefing the participant about how the data will be used and stored and how confidentiality/anonymity will be maintained. If the data are to be shared beyond the supervisor and student then explicit consent should be obtained. All written records should be stored in a secured environment.

Impartiality

Researchers should be honest. Any conflicts of interest need to be declared and dealt with. This is of importance with insider research or where research involves colleagues.

Data protection

Information needs to be given to the participants about the nature of the data to be collected, how this will be stored and for how long. Specific mention of when and how the raw data (tapes, transcripts) will be destroyed should also be made.

Information is deemed personal if it can be used to identify an individual; for example:

- name;
- any unique ID (driving licence number, student ID number, employee number);
- physical description;

- 'enough' characteristics (e.g. within a small team, age may be enough to identify an individual).

Data are seen as sensitive if there is information about, for example, race/ethnicity; political belief; religion; physical/mental health or conditions (disability); sexual life; and criminal offences (actual or alleged). Different participants may see some requests for data as being sensitive to them. Whether an item is sensitive or not is determined from the participant's viewpoint, not the researcher's. Examples may include questions on their private life or their salary; and opinions which could cause them harm or embarrassment;

Information collected and processed should:

- be used for the stated purpose of the research;
- be accurate;
- be relevant;
- not be excessive (only information of relevance should be collected);
- be protected against unauthorised access;
- not be kept for longer than necessary (don't destroy too early in case the data are needed for supervisor scrutiny).

Any information with participant identifying information should be stored securely and away from the collected data. As noted above, this may include the name of individual; their position within a named company; or descriptors which would allow their identification within a small group sample. Where questionnaire data include identifying information (e.g. an email address recorded to enable a follow-up interview/survey or to provide debriefing information), this should be treated in the same way as interview transcripts and held separately and securely from the anonymised data.

Presenting data

Excerpts of transcribed interviews may be used in the dissertation in an anonymised format using a key (such as 'I01', 'Participant 1' or 'Interviewee A'). Similarly, field notes from observational studies can be included as long as these are anonymised. Any information that potentially might identify a particular participant should be removed.

Academic misconduct

Research ethics also extend to the full work presented by the student. Unacceptable conduct can include any of the aspects in Table 7.1.

Outline ethical review

Most institutions have their own ethical policy and provide the relevant documentation to support the research process at undergraduate or postgraduate level.

Table 7.1 Unacceptable conduct

Type of conduct	Description
Fabrication	Making up any aspect of data or supporting evidence
Falsification	Presentation of data or findings which have been altered
Plagiarism	Using the work of others without acknowledgement
Misrepresentation	Presenting a flawed version of the data
Mismanagement	Failing to keep accurate records of the research procedures followed and the data obtained
Breach of duty of care	Deliberately or negligently giving the identity of individuals involved in the research without their consent Placing the researcher or others in danger of harm Failing to protect the environment
Dishonest practice	Actual or attempted bribery Submission of a thesis produced by another person

Table 7.2 gives an indication of the aspects which would usually be covered in an ethical review and demonstrates the relationship between the ethical review process and the process of research.

Participant information sheets and consent forms

An information sheet should be produced for each participant. This should include:

- the title of project;
- an opening statement explaining what the study is about, why participation has been invited and what this will involve (how much participation, how many times, where, how long);
- a statement of how the data will be collected and stored, and for how long;
- details of who will have access to the data, what will happen to it when the study is over and what outputs will result (dissertation, publications, presentations, journal articles);
- a statement of confidentiality or anonymity with details of any systems to protect the participant;
- information about how participants will be debriefed;
- notice of the voluntary nature of participation and the possibility of withdrawal of consent, and how this can be done;
- the student's contact details – students are advised not to give their home address or home telephone number as points of contact, but to use the address of the institution and student email address;
- details of who else can be contacted with questions or concerns about the study.

A documentary record must be kept, including some form of indication that each participant has been briefed about the study and understands what is entailed. A

Table 7.2 Example sections in an ethical review

Section in ethical review	Considerations/sections of the book to refer to
Title of project	'The dissertation title' in Chapter 3
Aims and objectives/hypothesis	'The hypothesis and null hypothesis' and 'Aims and objectives' in Chapter 3
Overview of the topic area	'The literature review' in Chapter 5 Demonstrate key aspects of the literature that have informed the direction of the research
Research design	Chapter 6 'Research concepts' Demonstrate the quality of the research design and its suitability, showing that participant time is not being wasted
Supporting documentation	This includes, for example, letters to participants, briefing documents and consent forms
Type of data analysis	Chapter 8 'Data analysis and presentation'
Any possible negative consequences to research participants and how these are to be limited	Consider informed consent, integrity, confidentiality/anonymity, impartiality, nature of the data, data protection, and storage Where information is withheld from participants or any aspect of the research requires misleading participants, detailed justification of why this is necessary should be given
Participant rights and how participants will be informed of these	Informed consent Right to withhold information Right to withdraw Contact point for complaints
Participant debriefing	Participants should be informed of how they can find out about results Commonly, an offer is made to share the research results with interview participants (e.g. sending a summary via email); or if results of the research are publicly available, details should be given of when and how these can be accessed Ensure there are no anonymity and confidentiality conflicts
Conflict of interest	If there are no conflicts of interest, declare that this is the case; or identify those that do exist

common way of recording this information is to ask participants to sign a copy of the consent form, which should include confirmation of the following:

- that they have had all the information they need;
- that they understand their ability to withdraw from the study (as applicable);
- that they understand the level of anonymity applied to responses;
- that they agree to take part.

Both the participant and the researcher should each sign two copies. The participant receives one copy. The second copy is for student record.

This contains identifying information. Do not place signed versions of this sheet in your dissertation but keep securely as part of the audit trail.

Sources and further reading

BIS (2007) *Rigour, Respect, Responsibility: A universal ethical code for scientists.* [Online]. Department for Business, Innovation & Skills and Government Office for Science (12 September). Available from: https://www.gov.uk/government/publications/ universal-ethical-code-for-scientists (accessed 1 August, 2015).

ESF (2011) *European Code of Conduct for Research Integrity.* [Online]. 4 March, 2011. Available from: www.esf.org/fileadmin/Public_documents/Publications/Code_ Conduct_ResearchIntegrity.pdf (accessed 1 August, 2015).

RCUK (2013) *RCUK Policy and Guidelines on Governance of Good Research Conduct.* [Online]. Updated July 2015. Available from: www.rcuk.ac.uk/Publications/researchers/ grc/ (accessed 1 August, 2015).

UNIVERSITIES UK (2012) *The Concordat to Support Research Integrity.* [Online]. Available from: www.universitiesuk.ac.uk/highereducation/Documents/2012/ TheConcordatToSupportResearchIntegrity.pdf (accessed 1 August, 2015).

WCRI (2010) Singapore Statement on Research Integrity. [Online]. *World Conference on Research Integrity.* Posted 22 September. Available from: www.singaporestatement.org/ (accessed 1 August, 2015).

Researcher question
Ethics

Will your research require consideration of ethical issues due to topic, methods selected or other factors?

Yes. I will need to undertake an ethical review of my work and will investigate the procedures used at my institution.

No. I am not sure that the nature of my topic will require a detailed ethical review; however, I recognise that many of the aspects outlined will need to be considered in order to improve the work. I realise that if my topic or methods change, I may need to revisit the need for ethical approval to the requirements of my home institution.

Health and safety

Researchers have a responsibility to ensure that their own health and safety and that of others is not compromised by their actions (or lack of action). Risk assessment must be undertaken before data collection takes place, and no student should undertake any activity that puts them or others at unreasonable risk. The sort of issues that will come up will depend on the nature of the project but may include consideration of:

- where the data collection will take place;
- how personal safety can be managed;

- any specific concerns;
- how these risks can be managed.

Health and safety in laboratory-based research

In the assessment of risk associated with conducting work in the laboratory, the student should give thought to:

- potential hazards (things that may cause harm) during the work;
- the level of risk (the chance that a hazard could cause harm and the amount of harm this could cause).

Safe working is always a priority, and the student must follow protocols that are in place for laboratory-based work. Where necessary, the student should use protective equipment. Reporting of near misses or incidents is good practice so that potential hazards are passed on for others using the laboratory.

Health and safety in field-based research

Undertaking research in the field involves other sorts of risk, including those that are related to known existing hazards, for which steps can be taken in advance to mitigate risk, and chance events that are unpredictable. This will almost certainly include assessment associated with:

- the fieldwork itself – personal safety and security;
- the fieldwork site or location – for example, any significant natural hazards or health risks and levels of political threat;
- travel to, around, and from the location;
- the knowledge and experience of the participants;
- any medical conditions that the researcher has;
- emergency or contingency plans.

If the research is off-campus, key questions to be addressed include: How will you travel to and from the data collection venue(s)? How will personal safety be maintained when travelling to and from the venue(s) and during periods of data collection?

Good practice guidelines are:

- Outline any potential risks associated with the venue/location/topic.
- Leave the plan for data collection itinerary for the day with a trusted individual (to include the time you are leaving, the venue, anticipated time of return, and mode of transport).
- Leave contact details (a mobile phone number and/or the contact number for the venue).
- Check out and check in with the trusted individual.
- Provide instructions on what to do if you do not check in with your trusted individual.

While these are general points for the planning of field-based data collection, there are additional concerns associated with different types of location. These are outlined in Table 7.3.

Table 7.3 Location-based hazards

Activity	Possible hazards to be considered (this list is not exhaustive)
Any outdoor activity	Climate conditions (heat/cold/fog/rain/sunlight/snow/wind)
	Trips/slips/falls
	Biting/stinging insects
	Dehydration
	Getting lost
	Manual handling of equipment
General	Travel to the field location
	Illness
	Food poisoning
	Communication difficulties (e.g. no signal for mobile phones or phones running out of battery power)
Mountainous areas	Altitude
	Objects falling from height
	Communication difficulties
	Access to emergency services
	Natural hazards (e.g. landslips)
	Access to food/water/safe refuge
Rivers or open water	Drowning
	Pollution
	Waterborne diseases (e.g. Weil's disease)
	Bad ground conditions (e.g. quicksand, mudflats)
	Wildlife (e.g. wasps, mosquitoes)
Coastal areas	Tides
	Natural hazards (e.g. rock falls, storm waves)
	Bad ground conditions (e.g. quicksand, mudflats)
	Wildlife (e.g. jellyfish)
Agricultural areas	Natural hazards (e.g. thunderstorms, flooding)
	Vehicles/farm machinery
	Domestic animals (including animals with young)
	Wild or feral animals
	Falls into depth or from height
	Possible diseases (e.g. Lyme disease, Weil's disease)
	Chemicals (e.g. herbicide or pesticide spray)
	Game shooting
	Landowners (always ask for permission for entry to private land)
Urban areas	Isolation/lone working
	Traffic including commercial vehicles
	Abuse or physical violence/becoming a victim of crime/uncontrolled dogs
	Natural hazards (e.g. flooding)
	Dangerous structures
	Chemical, biological and radiological hazards (current or historical)

Table 7.3 (Continued)

Activity	Possible hazards to be considered (this list is not exhaustive)
Construction sites	Falls from height
	Falling objects
	Machinery including plant/traffic
	Dangerous structures
	Trenches
Wilderness areas/nature reserves	Wild animals (especially animals with young)
	Water purity
	Possible diseases (e.g. Lyme disease, Weil's disease)
	Game shooting
	Access to emergency services
	Falls from height (e.g. in old quarries or where natural features might lead to falls)
	Fire
Research in areas outside the UK	Cultural sensitivities generally and, in particular, around the fieldwork activities
	Political instability and/or local hostility
	Natural hazards
	Access to emergency services
	Water purity
	Tropical disease
	Venomous snakes and insects

Ultimately consideration of these issues should allow appropriate controls or safety precautions to reduce the likelihood or severity of the identified risks. Table 7.4 suggests ways that hazards related to fieldwork can be reduced.

In general, aspects to check include:

- the project and the expected outcomes from the fieldwork are clearly defined;
- health and safety guidelines of the home institute have been followed;
- insurance has been taken out where necessary;
- details of planning and logistics have been passed on, including information about travel arrangements and a detailed itinerary where appropriate;
- there is a plan for action in an emergency (especially if working alone or where there are pre-existing medical conditions), which may include how to communicate in the field location.

For fieldwork outside the UK, the government's online advice on foreign travel provides useful information for travel to different countries (https://www.gov.uk/foreign-travel-advice). This includes details of any requirements or recommendations for immunisation. Advice on immunisation or vaccinations, travel FAQs and other guidelines are also available on the NHS website (www.nhs.uk/). Visa requirements must be checked if undertaking fieldwork abroad.

Where the nature of the fieldwork could expose participants to pathogens, such as work with soil or animals, immunisation, particularly for tetanus, is strongly recommended.

Table 7.4 General hazards and lowering risk

Hazard	Some suggestions to lower risk
Trips/slips/falls	Take care. Wear appropriate clothing for the activity (e.g. sensible boots or shoes, long trousers, gloves).
Objects falling from height	Wear a hard hat.
Falling from height	Be aware of sudden drops, particularly in strong winds.
Collisions with vehicles	Wear a high-visibility vest.
Weather conditions	Hot/sunny – wear a hat, apply sunscreen, and carry water. Cold/wet – wear multiple layers, dress warmly, carry spare dry clothing in a bag. Be aware of symptoms of hypothermia. Do not work in heavy rain conditions.
Physical terrain	In coastal areas, check tide times and consult maps to avoid being cut off. Be aware of mudflats and possible landslips from cliffs. When working with streams, be aware of changing water levels. Take care of footing in slippery conditions.
Animal or insect bites or stings	Treat as soon as possible and seek medical advice if necessary.
Cultural sensitivities	Be aware of differences in relation to acceptable behaviour and dress in different environments.
Getting lost	Carry a paper map and compass and know how to use them. Do not rely only on technology.

Lone working should be avoided where practicable. Where lone working is unavoidable, relevant risk assessment must be carried out and measures put in place to ensure that the work can be carried out as safely as possible.

If a student has any doubts about undertaking research in a particular location, the area should be marked as 'inaccessible' and they should not proceed.

There should be a clear plan of action to deal with any emergencies, such as accidents. As a minimum, this might include: knowing how to contact the emergency services, particularly overseas; location of the nearest hospital; means of communication – this may be a mobile telephone or walkie-talkie.

Students are advised to seek a basic understanding of first aid. Fast first aid tips are given on the Red Cross website (www.redcross.org.uk/) or consider doing a course with St John's Ambulance (www.sja.org.uk/). A recommendation is made to carry a personal first aid kit into the field (e.g. the Lifesystems range or items available from St John's Ambulance online store).

Summary – assessing and mitigating the risks

The procedures and policies of the home institution and guidance from supervisor or technical specialists should always be sought. In the absence of specific university guidance, a risk assessment may include consideration of items in Table 7.5 as a starting point.

Table 7.5 Considerations in risk assessments

Description of the fieldwork	The activities planned, the purpose, the outcome
Itinerary	Dates and locations
	Travel arrangements
	Nature of accommodation (campsite, youth hostel, hotel) and contact addresses
Student contact details	Mobile number; email address
Further contact details	E.g. a family member, fieldwork 'buddy', a friend who will have an idea about the student's location if working off-campus
Nature of site(s) to be visited/used	E.g. laboratory, office, workshop, construction site, remote area, city centre
Processes involved	Procedures to be used
	Manual handling
	Operating machinery or specialist equipment
Transport	To and from location(s)
Violence	Potential for violence in location
Cultural considerations	Consider those specific to the setting as well as participants
Individual circumstances	Medical conditions
Other hazards and the measures taken to reduce these	Consider general and location-based hazards as outlined in Tables 7.3 and 7.4
Legal requirements	Access to the area to do fieldwork must meet legal requirements – ensure that any external permissions needed (permits, permissions from landowner, occupier or relevant organisations) are identified and sought
Emergency procedures	Plan for any sorts of emergencies that can be foreseen
	Ensure contact details provided are up to date
	Carry details of medical conditions

The health and safety office or nominated person within an institution must be consulted to ensure that all necessary assessment has been done and that the student has approval to collect data. A copy of the completed risk assessment for the research should be given to the supervisor and appended to the dissertation.

Further reading

BRITISH RED CROSS (2015) First Aid. [Online]. Available from: www.redcross.org.uk/What-we-do/First-aid (accessed 24 August, 2015).

HSE (2015) Controlling the Risks in the Workplace. [Online]. The Health and Safety Executive. Available from: www.hse.gov.uk/risk/controlling-risks.htm (accessed 1 September, 2015).

NHS DIRECT (2014) Insect Bites and Stings – Prevention. [Online]. Page last reviewed: 27/06/2014. Available from: www.nhs.uk/Conditions/Bites-insect/Pages/Prevention.aspx (accessed 24 August, 2015).

ST JOHN'S AMBULANCE (2015) Home page. [Online]. Available from http://www.sja.org.uk/ (accessed 24 August, 2015).

Researcher question
Health and safety

Do you need to consider the health and safety implications of your research work?

Yes. My work will require me to produce appropriate documentation on the likely hazards and to assess how the impact of these may be reduced in order to minimise the risk to myself and others. I undertake to seek specialist support where needed.

No. The nature of my work does not have additional health and safety risks. I realise that if my topic or methods change, I may need to revisit the need for risk assessment to the requirements of my home institution.

METHODS OF DATA COLLECTION

This part of the chapter builds on the discussion of methods in Chapter 6, covering in more detail the data collection methods commonly used in student dissertations. Students must choose between a varied range of possible data collection methods, each with their own advantages and disadvantages. The key thing is to make sure that the data that will be gathered fits with the aims and objectives of the research. Students are advised to consider the following general points when approaching *any* method of data collection:

- Should the problem be researched using this method?
- Will the method contribute to achieving all of the objectives?
- Does the literature provide any guidance for using this method?
- What type of data do you want to collect?
- How much of this type of data is needed to meet the aims of the research? Is this level possible given the resources available?
- For research involving people, who is to be approached and how will respondents be identified?

Interviews

Interviews are used to collect detailed information from a relatively small number of individuals. These are often used in undergraduate dissertations as they allow a detailed understanding of a topic area from a selected point of view. Anonymity is not usually possible as the student knows the respondent's details; however, data can be reported anonymously. Interviews may bring to light information which could not readily be discovered from any other means. The interview process allows for clarification of points and probing questions to be used. Interview responses may be subjective in nature and some skill is needed to analyse these well. The low number of interviewees means that making generalisations is hardly ever appropriate as the sample is unlikely to be representative of the underlying population. Interviews are traditionally conducted face-to-face, but may be undertaken by Skype, by telephone or by extended email correspondence.

Like all methods of data collection, interviews have some limitations:

- The skill of the interviewer is a significant factor in the quality of data available.
- Unintended bias can result from the sample interviewed.
- Misinterpretation of written transcript data is possible unless notes are taken on body language and tone.
- Where interviewees are reluctant to be recorded and notes are taken instead during an interview, this can be challenging (e.g. for individuals who do not use shorthand).
- Arranging and undertaking interviews can be time-consuming.
- Travelling to locations may have cost implications.
- Respondents may give the responses they think you want, or need, to have rather than stating what they genuinely believe.

The limitations of sample size (having a small number of interviewees) makes the consideration of the sampling design critical (see 'Sampling' in Chapter 6).

Doing interviews

As shown in Table 7.6, there are different ways of carrying out interviews, and these will have a bearing on construction of suitable questions.

Question writing

After choosing to conduct either fully structured or semi-structured interviews, questions can be formulated. (If unstructured interviews are being carried out, a list of questions is not appropriate.) A three-stage approach to developing an interview schedule is recommended, as outlined below.

STAGE 1: IDENTIFY THE INITIAL QUESTION SET

With reference to the literature reviewed and the initial proposal, create the initial list of questions. A fully structured interview will require many more questions than a semi-structured interview.

1. List all items about which information is required.
2. Group these into themes or sets.
3. Write the first draft of the questions.
4. Check the wording of each question carefully.
5. Examine the list critically and remove unnecessary questions.

Table 7.6 Forms of interview

Type of interview	Characteristics	Advantages	Disadvantages
Focus groups/ group interviews	Several people are involved at once	Can be less expensive than several solo interviews Being in a group situation can encourage contributions that would not happen otherwise A large amount of data can be gathered quickly	Difficult to record Ideally two people are needed, one to ask questions and one to take notes Group dynamics can affect the results The number of questions has to be limited Audio recording can be difficult to analyse retrospectively
Telephone interview	The interview process is carried out without face-to-face meeting	Very cost-effective Easy to organise as no venue needs be booked Lack of visual information may reduce unconscious interviewer bias	Breaking the ice can be much more difficult Loss of body language and non-verbal information Recording needs to be considered (one option is to use a speaker) Additional care is needed with informed consent and the audit trail Typically of shorter duration than face-to-face interviews and there may be reduced depth of information
Email interview/ iterative text conversation	An extended conversation where responses to the questions are probed through further text-based questioning	Cost-effective No travel time or venue constraints Interviews can be carried out concurrently Opportunity for interviewees to reflect on their initial responses and provide clarity as conversation progresses Recording occurs as a matter of course	Care must be taken to make this method an 'interview' rather than a 'letter' Responses must be probed to gain maximum detail Loss of non-verbal information Process may need several weeks of communication/ dealing with increasingly lengthy responses Ethical issues need to be carefully considered

Any data not needed by the research, particularly relating to the respondent's private life, should not be sought.

STAGE 2: FORMULATE THE DRAFT INTERVIEW SCRIPT

1. Questions should be placed in a logical order, usually moving from simple questions through to those that ask for more detailed information. Careful

review of the questions generated by the first stage is now needed. These should not be leading or confusing, too long or too complicated.

2. Check the overall length of the interview schedule. Having too many questions is risky and could cause interviews to overrun or result in failure to collect all the relevant or required information.

STAGE 3: FORMULATE THE FINAL SCRIPT

1. Finalise the questions.
2. Add in any protocol prompts needed.
3. Pilot the interview to check that it is feasible within the time limits.
4. Additional reserve questions may be created in case of running under time, or to probe more deeply if the time allows.
5. Consider this data in terms of how it will be analysed.
6. Adjust if necessary and re-pilot.

Practical considerations

Careful planning is needed before, during, and after the interview in order to ensure that the process runs smoothly.

BEFORE THE INTERVIEW

Guidance for planning in advance of carrying out interviews:

1. Identify the format of the interview (Table 7.6).
2. Create a schedule with interviewer prompts as required.
3. Determine what recording methods are appropriate – participant permission will be needed for audio or video recording.
4. Identify interviewees.
5. Arrange the date and time for interview, giving the expected duration (e.g. between 30 minutes and an hour).
6. Arrange the location for the interview (somewhere quiet with informal seating).
7. Organise the approach to obtaining informed consent.
8. Consider provision of information prior to interview – consider if the participant will be sent a copy of the questions beforehand.

DURING THE INTERVIEW

An example protocol for carrying out an interview is given in Table 7.7. In addition, remember the following points:

- Be polite and courteous.
- Be interested and alert.

Table 7.7 Example interview protocol

Protocol	Notes
Interviewee	Use a code here, and record the interviewee's details in a separate place
Date	Record information
Time of interview	Record information
Place	If this could identify a participant, use a code and record full information in a separate place
Position/role of interviewee	If this could identify a participant, record information in a separate place
Provide an overview of the project	Giving a very brief summary verbally is a good way of settling into the interview and reassuring the participant. E.g.: 'Thank you for agreeing to participate in this study on ... [title] I am undertaking [number] interviews with [sample population] which will be used to answer [research questions]. All interview data are recorded without participant names and will be referred to by a participant code. A transcript will be sent to you within two weeks for you to review. The interview will take between [shortest and longest duration expected]. Do you have any questions?' This information should also be given in detail on the request for interview/participant information sheet and/or consent form
Gain participant consent	Make sure you have permission to record Have the interviewee read and sign the consent form
Do the interview	Take notes on paper throughout in case the audio recording fails for any reason
Close the interview	Thank the interviewee Reassure them of the methods to be used for data protection Confirm debriefing

- Do not judge the participant's answers as right or wrong.
- Ensure, as the interviewer, that you fully understand their answer.
- Note non-verbal information (doubt, discomfort, confidence).
- Keep on track (topic and time).

AFTER THE INTERVIEW

Every interview needs to have a clear audit trail of consent forms and notes or transcripts. This is important in case the supervisor has queries about what data were collected or if you are required to prove that your analysis is appropriate. Notes should be written up as soon as possible while the interview is fresh in the researcher's mind so that the data are as accurate as possible.

Transcription may take a considerable time. One hour of audio may take anything up to six or seven hours to transcribe. If this is unmanageable, an audio typist can be hired, but this will need to be incorporated into the ethical review and the participant consent.

Sending a copy of the transcript to a participant ensures that their opinions have been correctly represented. This may need to be handled sensitively; for example, by removing all identifying information. The text version of an interview will sometimes read very differently to the interview as experienced. For instance, information given with a tone of sarcasm or in jest will not necessarily be recorded as such. Non-verbal information can give context in these instances and may need to be annotated on the script.

Overcoming the limitations

No research method is without flaws but the limitations of the interview can largely be dealt with by careful planning. Key points to address are:

- Ensure that the participant is fully briefed about the expectations of their participation before the interview.
- Identify and mitigate ethical concerns.
- Identify means of analysis early in the process.
- Pilot the interview in order to make sure the questions make sense and can be answered in an appropriate time frame.
- Be aware of the specific limitations of the method selected.
- Discuss concerns with the supervisor.
- Ensure participant debriefing occurs.
- Keep records up to date to evidence actions.
- Keep data secure.

Researcher question
Interviews

Do you intend to use interviews as part of your primary data acquisition?

Yes. I intend to use interviews and I am clear on the requirements of the process in order to undertake these to an appropriate standard. I understand that I may also need to consider ethical (and health and safety) requirements.

No. The nature of my work does not lend itself to data collection by interview. I realise that if my topic or methods change, I may need to revisit this section.

Questionnaires

Questionnaires are widely used in social research methods as they offer the advantage of a standard design that can be used to collect a small amount of data from a wide population. The method is relatively low cost, especially where the questionnaire is administered electronically. Anonymity is possible but not recommended unless required for ethical reasons. Disadvantages of this relate to uncertainty about the identity of the respondents, their motivations in completing the questionnaire and the truthfulness of their responses. There is often industry fatigue due to the volume of student questionnaires in circulation, and this can reduce response rates. Other factors may also act to reduce response rates to a level lower than anticipated by the researcher, and poor response poses problems in interpreting the data that are gathered. The data may, therefore, be of poor quality or unrepresentative of the underlying population.

Students are advised to consider the following when using a questionnaire as part of dissertation research:

- Will the questionnaire help to achieve the purpose of the research?
- What data will need to be gathered?
- Over what time period will data be collected?
- It is important to consider the sampling strategy early on, especially if the aim is to select a sample that is representative of the target population.
  - What is the population of interest and how will the sample be selected?
  - How many respondents are needed?
- A questionnaire cannot be clarified or amended once sent out, so sufficient time needs to be built in for design and piloting.

The limitations of questionnaire-based research include:

- The answers given by a participant may be ambiguous, or responses may be received in a format which was different to the one designed by the researcher.
- Respondents may give the responses they think a student wants or needs rather than what they genuinely believe to be true.
- With self-completion questionnaires, the respondent will typically see some, or all, of the questions before starting to complete the questionnaire, which might influence their responses.
- With electronic questionnaires, the respondent may lose interest or Internet connection while completing the form.
- The person intended to receive a self-completion questionnaire may not be the one who completes it.
- Cost and time requirements must be identified early in the process.

Construction of a questionnaire

Writing questions for a questionnaire is an art form, and most students will not have a lot of experience in designing a questionnaire. Slight changes in how a question is phrased or the format used to collect data can have a significant effect on respondents' answers. Some general guidelines follow:

- Clearly state the purpose of the questionnaire.
- Care is needed in the construction of questions, especially where the questionnaire is not completed in the presence of the researcher – questions should
  - be simple and carefully worded, not vague or ambiguous
  - be objective and unbiased and not lead the respondent
  - take account of potential respondent sensitivity.
- The questionnaire should provide the researcher with data that can be meaningfully used in research; consider the suitability of different response formats.
- Give consideration to the overall presentation and layout of the questionnaire
  - use a clear and attractive layout with logical question progression
  - questions must be accompanied by clear and simple instructions where necessary.
- It should be long enough to provide sufficient data but not so long as to be onerous to the participant.

The following three-stage process is suggested for questionnaire construction.

STAGE 1: IDENTIFY THE INITIAL QUESTION SET

With reference to the literature review and the initial proposal, create the initial list of questions. Try to keep these in approximate themes, as drawn from the literature, although the order is not vital at this stage.

1. List all items about which information is required.
2. Ensure that all information requested has a purpose related to the research outcomes.
3. Group the items into general themes.
4. Write a first draft of the questions.
5. Examine the list critically and remove questions that will not provide data relevant to the study, particularly where these relate to personal information.

STAGE 2: REFINE THE QUESTIONS AND LAYOUT

1. Decide on question type – open or closed (refer to 'Question types' below). Try to use closed questions wherever possible.

2.   Choose appropriate response formats for closed questions (refer to 'Question types' below). If responses are provided for respondents to choose from, make sure the options are comprehensive (or include an 'other' field).
3.   Are questions concise and the wording clear with minimal use of jargon?
     a.   double negatives should be avoided
     b.   add instructions where needed (e.g. 'Choose ONE of the following options')
     c.   check that each question deals with one query – do not combine questions
     d.   questions should not be leading
     e.   the questionnaire should not be a quiz – do not ask knowledge-based questions
     f.   abbreviations or technical terms should be avoided (unless these are pivotal to the research question and it is known that the respondent will understand these).
4.   Consider the logic and clarity of the layout
     a.   sort questions into logical order – this usually works best going from closed questions to open-ended questions
     b.   don't put the most important question(s) last
     c.   direct the respondent to appropriate questions where necessary (e.g. 'If you answer no, go to question *x*').
5.   Consider overall appearance and length
     a.   try to fit questions on two sides of A4 if possible as lots of pages may be off-putting to respondents.

STAGE 3: CHECK THAT THE QUESTIONNAIRE WILL PRODUCE SUITABLE DATA

1.   Consider how the data will be analysed and presented.
2.   Pilot the questionnaire – look for ambiguities and make changes as required.
3.   Based on the piloting process, adjust question type or wording and questionnaire length and layout
4.   Re-pilot if necessary.

Question types

Questions may be open or closed. In an open question, the respondent is allowed to give the answer as they choose, using their own words. This type of question is useful where it is not possible to predetermine responses, but when it comes to analysing the data collected, the answers will likely need to be coded. This can be time-consuming and is something researchers often prefer to avoid when using questionnaires.

   In a closed question, the type of response is determined by the researcher. Care must be taken to ensure that all the answers that a participant could use are listed; it is possible to include a response such as 'Other – please specify', but try not to rely heavily on this as the data will need to be coded (in the same way as responses to open questions) later on.

Closed question response formats

There is a variety of response formats to choose from when using closed questions. These include: dichotomous response (yes/no); a single number indicating frequency (e.g. total number of employees at a company); choosing from ranges or bands (e.g. 'less than 10 employees', '10 to 20 employees', '20 to 30 employees', etc.); checklists; rating scales; and numerical ranking; as well as open-ended responses where respondents have been offered an 'other' category. Further description is given below for checklists, rating scales and numerical ranking.

CHECKLISTS

In this response format, state how many choices should be made and the criteria for responding: for example, 'tick one answer' (Table 7.8); 'tick the most important'; or 'tick all that apply'.

The checklist format can be presented as a grid with numerous statements. Table 7.9 illustrates this format, asking for one selection per row. Depending on the question being asked, it would also be possible for the respondent to tick 'all that apply'.

Table 7.8 Checklist method – one selection

Which of the following responses do you think most represents the current situation? *Tick only one.*

Question response 1	☐
Question response 2	☐
Question response 3	☐

Table 7.9 Checklist method – grid

In the box below, a number of statements are provided. For each statement, please tick the option (A, B or C) that you feel fits best.

Criteria	Description of options		
	Option A	Option B	Option C
Statement 1			
Statement 2			
Statement 3			
etc.			

RATING SCALES

Rating scales are popular and a numerical value can be allocated to the data to allow statistical analysis (ordinal data). The Likert-type scale, commonly used to measure attitudes, is a rating scale in which the respondent chooses between ordered categories (e.g. 'strongly agree' to 'strongly disagree').

The number of points on a scale is an important decision. If there are too many, the respondents will struggle to provide an acurate rating; too few and the data will not give the specificity needed for analysis. Odd numbers of categories (e.g. five or seven) allow for the collection of 'neutral' responses in the middle of the scale, whereas even numbers of categories (e.g. four or six) force the respondent to choose one or other side of the neutral area.

Responses are typically arranged horizontally. Table 7.10 shows how this can be presented for a single statement while several statements are presented in grid form in Table 7.11. Responses may also be arranged vertically (Table 7.12).

Rating scales come in a variety of forms. Care is needed to inform the respondent exactly how they should fill in the table (see Tables 7.13 and 7.14). An issue that may arise when this method is used involves respondents choosing the same rating across a range of statements, perhaps because of questionnaire fatigue.

Table 7.10 Likert scale – horizontal

	Strongly agree	Agree	Neutral	Disagree	Strongly disagree
Statement					

Table 7.11 Likert scale – horizontal grid

		Strongly agree	Agree	Neither agree nor disagree	Disagree	Strongly disagree
1	Statement 1					
2	Statement 2					
	etc.					

Table 7.12 Likert scale – vertical

Statement	Please tick
Too little	
About right	
Too much	

Table 7.13 Numerical rating (1–5)

On a scale of 1 (strongly disagree) to 5 (strongly agree), please rate each of the following statements:

	Score
Statement 1	
Statement 2	
etc.	

Table 7.14 Numerical rating (score out of 10)

Give a rating out of 10 for each of the following statements, where 10 is of high importance.

Criteria	Rating
Statement 1	/10
Statement 2	/10
Statement 3	/10
(etc.)	

NUMERICAL RANKING METHOD

This is similar to the rating method, except that the respondent is asked to rank a set of statements in order indicating their importance, priorities or preferences. The ranking scale depends on the number of statements to be placed in order (e.g. in Table 7.15, there are eight statements so ranking is from 1 to 8).

Increasing response rates

Low response rates are usually the main concern of students using questionnaires to gather data. While response rates in excess of 60 per cent do occur, response rates may be as low as 10 per cent. This was referred to in Chapter 6, and Table 6.9 offers suggestions to increase response rates based on how the questionnaire is constructed. Others points to consider in terms of maximising response are discussed here.

Table 7.15 Numerical ranking method

Please place the following statements in order of priority starting with '1' to indicate the most important, '2' for the second most important, and so on.
Circle one number per row.
Select each number only once.

1	Statement 1	1	2	3	4	5	6	7	8
2	Statement 2	1	2	3	4	5	6	7	8
3	Statement 3	1	2	3	4	5	6	7	8
	(etc. to a total of 8 statements)	1	2	3	4	5	6	7	8

A covering letter (which may be a paragraph of text at the top of the questionnaire) can increase responses since this reassures the participants about the purpose of the survey and what they should expect as well as emphasising the useful contribution they are making to the study.

Make the process as easy as possible for the respondent. The survey should be easy to fill in and not too long. If sending postal questionnaires, the response rate can be significantly increased by enclosing a reply paid envelope (though this greatly increases the cost). Use pastel coloured paper so the survey does not easily get lost on desks.

Provide an incentive to respondents. This could be in the form of a charitable donation in return for participation, resources permitting, or the offer to share the research findings.

Consider what are the most relevant and convenient distribution means for the sample; for example, via online survey tools, using email, distribution of the questionnaire in person, distribution via a company mailing list, etc.

Following up initial distribution of questionnaires with a gentle reminder to return the completed questionnaire can be done by letter or telephone, as appropriate.

In general, being polite and courteous toward potential respondents can encourage their participation.

Overcoming the limitations of a questionnaire

Careful pre-planning and pre-testing of a questionnaire can significantly reduce some of the principle shortcomings. Students are advised to address the following when using the questionnaire method:

- Examine other questionnaires to identify possible clues.
- Allow for a piloting stage in order to get valuable responses and to detect areas of possible shortcomings.
- Obtain as much feedback as possible in order to make sure it is easily understood by different audiences.
- Trying different question orders at the pilot stage may assist in finding the best approach.
- Experiment with different types of questions, both open and closed.
- Some of the limitations of the questionnaire method can be overcome by supplementing it with personal interviews.
- Liaise closely with the supervisor and make sure that s/he has seen and approved the questionnaire.

Observation as a data collection method

'Observation' describes a range or continuum of data collection techniques. These are illustrated in Figure 7.1.

Researcher question
Questionnaire

Do you intend to use questionnaires as part of your primary data acquisition?

Yes. I intend to use questionnaires and I am clear on the requirements of the process in order to undertake these to an appropriate standard. I understand that I may also need to consider ethical (and health and safety) requirements.

No. The nature of my work does not lend itself to data collection by questionnaires. I realise that if my topic or methods change, I may need to revisit this section.

Observation is used as a data collection method in investigations based on both scientific and interpretive research perspectives. It is important to select the technique that aligns with the research perspective as well as the research aims objectives. It should also be borne in mind that the nature and characteristics of both the research setting and the researcher can determine the extent to which the observer could potentially fulfil the participant observation role.

Within the context of scientific research, observation serves to facilitate the evidencing and documenting of data measurements. Observation can also support the monitoring of the experimental environment and the variables therein. In this way, the experimental observer gathers evidence and documents data measurements as a means to monitor compliance of the experimental process. (See also the section below on 'Laboratory research'.)

Observations can form part of a quasi-experimental approach that examines the behaviour of groups in a real-world setting. Here, the non-participant observer is distant and separate from that which is being researched. It is essential that they do not influence, alter or impact the experiment in any way. An example of this is evaluation of the impact of refurbishment on different staff groups within an office building. Behaviour is observed before and after refurbishment. (See also the section below on 'Field experiments'.)

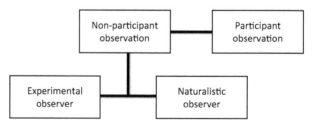

Figure 7.1 Continuum of observational roles

Observational data collection techniques for interpretive research encompasses naturalistic observation and participant observation. When undertaking interpretive observation, the researcher can be described as being the tool or instrument of data collection with the purpose of systematically collecting both verbal and non-verbal data. Within this perspective, the researcher-observer does not exert any control upon the researched or the research environment. Instead the observer seeks to gain an understanding of the setting.

In naturalistic observation, the researcher observes the social setting of the research environment at a distance and is 'other to' the social environment of the research setting. The naturalistic observer does not engage with the social setting in any meaningful way; instead, they simply observe and record verbal and non-verbal data. Webcams and other such modern-day technology can prove very helpful and useful in facilitating naturalistic observation within a geographically specific research setting.

Participant observation requires the researcher to gain entry to the social environment of the research setting (the 'field') and to establish and maintain a role within that environment. This can be potentially very challenging as social environments can be settings of great complexity.

Sociologists employ the metaphorical term 'actors' to describe people within social environments. The foundation for the use of this metaphor is that people ('actors') undertake to perform roles within social environments. This metaphor can prove very useful when seeking to observe social settings and the behaviours, conversations, rules and conventions therein. Social settings can be described as being made up of both 'front stage' and 'backstage' where the visible front stage can be openly observed, possibly by use of naturalistic observation, but the backstage is hidden and might only be observed and made accessible by participation within the social setting.

Portrayal of the role of the researcher and the purpose of the research

Before undertaking to observe a social setting, a researcher must decide the extent to which both their own role and the purpose of the research are to be presented to the research participants who will be observed. Not openly declaring your observational role and the purpose of the research can prove to be ethically challenging and may prevent the research being carried out.

When portraying the purpose of the observation, a number of options are potentially available, but only one is realistically possible within the context of a dissertation research project:

- *full explanation* (an overt approach) – research participants are openly informed of the research purpose (high ethical integrity – this option is what would commonly be expected within a dissertation research project);
- *partial explanation* – rather than research participants being fully informed of the research purpose, only partial details are provided (low ethical integrity);

- *no explanation* (may involve a covert approach) – research participants are provided no information as to the research purpose (low ethical integrity);
- *false explanation* – research participants are actively deceived about the research purpose (low ethical integrity).

There are three possible options for portrayal of the observer's role to the research participants, but two of the three approaches present significant ethical challenges:

- *overt observation* – research participants are made aware of the fact that observation is being undertaken and are informed of the researcher's identity (high ethical integrity);
- *partially overt observation* – research participants are made aware of the fact that observation is being undertaken but are not informed of the researcher's identity (low ethical integrity);
- *covert observation* – research participants are not made aware of the fact that observation is being undertaken (low ethical integrity). Here, it is important to balance the ethical considerations with attempts to reduce observer effects. If deception is used, some level of informed consent is still needed. For instance, the participants might be fully debriefed after the event and given the opportunity to withdraw their data.

Structured or unstructured

Observation can be categorised as being either structured or unstructured. A researcher undertaking observation as a data collection technique should clearly understand which of these approaches will be appropriate to their research.

- *Structured observation* utilises a pre-drafted checklist, or observation schedule, to direct or focus the observation upon known behaviours and activities within the field of study. Experimental observation and sometimes naturalistic observation can favour the use of observation schedules.
- *Unstructured observation* is unrestricted in its focus within the field and is used in exploratory ethnographic studies of social environments. Unstructured observation is commonly used within participant observation approaches deployed within ethnographic studies.

Whether observations are structured or unstructured, it is essential that observation data are captured in a timely fashion. Collecting data involving observation requires notes on what can be seen or heard as well as interpretation of the data. Notes in the field may be short and will need to be expanded fully, shortly after the collection.

Modern technology can provide a potentially very powerful and effective means to record conversations (audio) and behaviour and activity (visual) within a social setting. Hand-held tablets and smartphones also provide a very accessible means for writing and recording digital notes within the research setting. Hand-written observational notes also remain a worthwhile option. A notebook or journal will allow raw data and emerging ideas to be recorded, referred to and interpreted as work progresses. This might involve generating codes themes or categories and these will almost certainly be modified as the work evolves. An index or a file system may need to be created in order to keep track of everything collected. Whatever the choice or choices of recording method, the data must be accessible for the researcher to review time and time again whilst carrying out analysis. The data should also be kept available for inspection by your research supervisor.

Researcher question
Observation as a data collection method

If you do not intend to use observation, go to the next section.

If you intend to use observation as a data collection method, have you decided what observational role you are to adopt, how you will get access to the field of study and how you are going to record your observations?

Yes. I understand what observational role I am going to utilise for data collection. I have also determined how I am going to access the field of study and I have determined how I am going to record the observational data.

No. I have not decided which observational role is to be adopted for my research, nor do I know how I am going to get access to the field of study or how I am going to record and document observations.

Field experiments

An experiment aims to study the effect of certain variables by deliberately manipulating them while keeping all other variables constant. A well-planned and well-executed experiment will give good results with a high degree of validity. Chapter 6 provided a general overview of experimental approaches, which can be based in the field or laboratory. The former is the focus of this section (laboratory-based experiments are discussed below).

For example, this may involve deliberately altering or creating something in the field environment and then observing either public reaction to this or physical reactions. While some variables in the field can be changed by the researcher, it is not possible to put in place the same level of controls which would be possible in a laboratory environment. Here, a 'quasi-experimental' design or 'natural experiment' is more appropriate. This approach is used where the aim of research is to observe phenomena or interactions in real-world circumstances (as opposed to the artificial conditions found in the laboratory). In addition, there are some projects that, by their nature, could not be feasibly brought into a laboratory environment.

Field experiments are common in disciplines such as Environmental Science or Geography, but this approach is also highly relevant for any discipline which requires information on a physical location or site, including Architecture, Building Surveying or Construction/Project Management. It may involve the study of a physical phenomenon, or observation of how humans react to a physical phenomenon.

Field experiments in these disciplines may involve:

- collecting data from the physical environment while working within that environment (e.g. monitoring temperatures within a building, measuring fluid flows);
- collecting samples from the physical environment and subsequently undertaking laboratory testing (e.g. collecting samples from a building for testing while replicating conditions by environmental simulation);
- collecting data about the physical environment and the interaction of people within it (observations of how people use a place);
- collecting data from the physical environment and recording the researcher's reactions to this.

In some circumstances, it is possible to assess the effect of a situation where a change is imposed by an external agency. This will commonly rely on the measurement of two groups, one with and one without the intervention or difference. Human participants are not usually made aware that they are taking part in an investigation as this would almost certainly bias the results. This leads to ethical considerations, however, as it is not possible to gain informed consent or to debrief participants. The ethical implications of this should be discussed with a supervisor.

Researcher question
Field experiments

Do you intend to use field experiments as part of your primary data acquisition?

Yes. I intend to use *field experiments* and I am clear on the requirements of the process in order to undertake these to an appropriate standard. I understand that I may also need to consider health and safety (and ethical) requirements.

No. The nature of my work does not lend itself to data collection by field experiment. I realise that if my topic or aims change, I may need to revisit this section.

Laboratory research

Laboratory research is usually subject-specific. As such, the guidance provided here is general, and there will be particular considerations that are relevant to different disciplines. Typically, the design of an experiment should include an overview of the apparatus (equipment or instruments) used to collect data and what information is to be collected. Undertaking research in the laboratory requires an understanding of test methods – what is measured and what are the positives and negatives of each test (and how the negatives may be reduced). It must be carefully planned with appropriate contingencies, which may include changes in the planned procedures if it becomes clear that there is an unexpected occurrence.

The stages of laboratory research may be summarised as:

1. planning the experiment
    a. experimental design (should align with outcomes)
    b. materials/participants being tested;
2. assessing the logistical requirements
    a. ensure that the proposed procedure is feasible;
3. preparing health and safety and risk assessments;
4. obtaining/creating the samples, creating specimens;
5. carrying out the testing and recording the data;
6. analysing data.

An example is given in Table 7.16 of the considerations involved in planning a laboratory-based experimental design for a study to establish the efficacy of a cleaning methodology on bricks. Note that if the cleaning method was tested on a building in the field (i.e. observing colour changes on the façade before and after cleaning), the researcher would have no control over variables such as climate or underlying differences in the materials. Therefore, testing specimens in the laboratory can be considered a better option. This approach would involve consideration of the questions in Table 7.16.

Good practice for working in a laboratory

It is essential that safe practice is used in the laboratory. Training for the use of specific methods or equipment as well as general safety training may be needed. Staff and students working in a lab have a responsibility for their own and others' health and safety.

The laboratory manager should appraise the proposed work. Work should always follow the specific guidance associated with good laboratory practice, which includes but is not limited to the following:

*   Plan the work.
*   Maintain records.

Table 7.16 Example: to establish the efficacy of a cleaning methodology on bricks

Question	Advice
Is there a standard methodology for cleaning this type of material?	Check the literature
How many cleaning products will be tested?	Consider feasibility, health and safety, literature
What help and advice can I obtain?	Speak to supervisor and/or technical specialist
What type and size of bricks will be used?	Consider material sourcing
What pre-classification or pre-testing will be undertaken?	The sample size can be derived after initial classification to reduce variability if this is seen as important
How many specimens will be used?	This may depend on the extent and nature of the testing proposed
What materials and time will be needed to collect the data?	Produce a Gantt chart and consider feasibility; adjust if necessary
What indications will there be of success or failure of the experiment?	Consider whether a re-run of the procedure will be required with the same or different materials
How will data be presented and analysed?	Look at literature and identify if there are any conventions or methods which seem effective Early consideration will save time in the analysis and writing-up phases
What are the possible contingency plans, including alternatives to lower the risk of experimental failure?	Consider whether there will be enough time to collect alternative data if the experiment fails, or if is there another way to recover the project

- Review practice.
- Adhere to health and safety guidelines.
- Behave sensibly (don't run).
- Wear appropriate clothing and safety equipment (remove any items which pose a risk such as dangling jewellery, tie back long hair, etc.).
- Use equipment or chemicals appropriately and for the purposes specified in the design.
- Ensure appropriate training has been received.
- Know the safe working procedures (what to do if there is a fire, location of first-aider, etc.).
- Protect yourself and others from hazards.
- Don't eat or drink in the lab (others may not have cleared up chemical spills).
- Keep the work space clear and clean.
- Report
  - spills
  - breakages
  - damaged or broken equipment
  - near misses or incidents that result in injuries.

Laboratory experiments on objects

In general a set of questions will inform the type of experiment; for example, 'What is the effect of *x* on *y*?' Here, *x* is the independent variable (the parameter which can be altered). The dependent variable (affected by any changes to *x*) is *y*. When testing objects, it is usual to have several experimental groups and one control group. Results from the experimental group are compared against the results from the control group. A control sample gives confidence that the results of changes to the dependent variable are due to manipulation of the independent variable and not due to random variations.

For some laboratory procedures involving objects, it is possible to observe, record and test items over a period of time without unduly affecting the item being studied. For other procedures, testing involves undertaking actions that render the item incapable of further study after that point. This is termed 'destructive testing'. Every opportunity should be taken to maximise the data collected regardless of whether non-destructive or destructive testing or a combination of these is used.

Sample creation specific to destructive lab testing

The number of samples to be collected or created will have to be established prior to commencing the procedure. Generic advice on this will vary according to the nature of the investigation, the properties being measured, the test methods used, the resources needed and the support required – all may affect the sample size. Testing which is destructive in nature will generally need more specimens than a method that allows monitoring and evaluation on multiple occasions.

Consider the specimens required for testing of materials that can only be assessed once because the testing destroys the specimen. Table 7.17 shows the calculation where the student considers the (highly questionable) testing of only one specimen in each condition. Here, only eight specimens would be needed. However, the loss of any one specimen would significantly undermine the validity of the test, and there is no way of internally monitoring variability.

Increasing the number of specimens to (a more acceptable) 5 in each set would require 40 specimens in total (Table 7.18). In addition, a researcher may want to have a few spare specimens in case of accidental damage.

If a second variable needs testing then this would require a further 25 specimens and additional specimens would again be needed for each destructive test cycle planned. In this way, the specimen creation process can soon become unwieldy. Consider at an early stage what is logistically possible.

Case studies

Case studies were discussed in Chapter 6 in terms of the investigation of an entity such as an organisation or an individual in order to study the interaction of processes. A case study is a systematic investigation, commonly assembled

Table 7.17 Grid of specimen creation for experimental testing, low validity

	Control	Variable 1
Destruct test at time0	1	1
Destruct test at time1	1	1
Destruct test at time2	1	1
Destruct test at time3	1	1
Samples to be made	4	4
TOTAL		8

Table 7.18 Grid of specimen creation for experimental testing, increased validity

	Control	Variable 1
Destruct test at time0	5	5
Destruct test at time1	5	5
Destruct test at time2	5	5
Destruct test at time3	5	5
Spare set	5	5
Samples to be made	25	25
TOTAL		50

Researcher question
Laboratory research

Do you intend to use laboratory research as part of your primary data acquisition?

Yes. I intend to use laboratory research and I am clear on the requirements of the process. I understand that I may also need to consider health and safety (and ethical) requirements.

No. The nature of my work does not lend itself to data collection by laboratory research. I realise that if my topic or aims change, I may need to revisit this section.

---

using a mixture of qualitative and quantitative methods. The focus on in-depth understanding of one case means that generalisation is not usually an aim of the research. However, a successful case study will provide the reader with a detailed and rich picture of processes, patterns and relationships which other methods cannot. Where resources allow, multiple case studies can be implemented in order to compare similar cases in different contexts; this enables lessons common to different contexts to be identified.

Data collection

The mechanisms for data collection as part of a case study approach are numer-ous and, as noted, may be of a quantitative or qualitative nature. They should be carefully chosen to provide the data necessary to meet the aims and objec-tives of the study. (It should be noted that except in special circumstances, a case study based on literature must always be reported as such, as it utilises secondary data.)

As illustration, consider a case study that focuses on interactions of people and places: the user experience of an art gallery. This might involve collection of quantitative data recording level of footfall in specific spaces, such as exhibi-tion space, an atrium or a café, at different times of day. This may be enriched by collecting qualitative data through observation of activities, facial expressions or body language or by undertaking interviews. Alternatively, questionnaires might be circulated asking participants to rate or provide comments on certain locations on a floor plan; or photographs or simulated 3D models could be used to gain reactions, such as sorting by preference giving positive or negative words or phrases.

Targets about data needed should be set before the logistics of studying build-ings in the field are even considered. Tables 7.19 to 7.21 offer an overview of considerations when carrying out case study research on a building.

Piloting and pre-testing

Recommendations have already been given to pre-test methods, but it is worth revisiting this as it is an aspect often neglected by undergraduate students. Failure

Table 7.19 Some considerations when planning to visit a building and its occupants as part of a case study

Planning and information gathering	Example considerations (not all will apply to all cases)
Research design	What data are required for the study?
	Methods of data collection
	Ethics review
Logistics	Travel/accommodation
	Cost of entry
	Other costs
	Permissions needed
	Interview arrangements
	Printing questionnaires
	Create data collection pro forma
	Means of recording data (camera/notebook/Dictaphone)
Desk study	Before visiting, try to assemble information which will help to maximise the useful information that can be gathered; for example, maps, records, promotional literature, etc.

Table 7.20 Example of data collection methods for a case study on the use of a selected building

Data acquisition	Available methods (one or several may be selected in each case)
Description of the physical building (ensure you have relevant permissions before starting)	Illustrations and sketching Photography Video Written notes Audio recording
Form of construction	Plans, construction details Textbooks, papers, visitor guides Observation of the building Interviews with building designer/constructor/occupier
How people react to the building	Observation recorded by notes or video recording Questionnaire to users Interviews with owner/manager

Table 7.21 An illustration of tasks to be undertaken post visit

Task	Detail
Record all data	File and label photographs Annotate maps or plans Type up observations, notes and key findings Transcribe interviews Analyse questionnaires Collate desk study information Record and store safely any identifying participant information
Student's reaction to the work	Ensure there is a record of how the researcher reacted to the building
Data presentation	Create the visualisation for the dissertation (plans, sections, elevations, interior or exterior views, concept sketch diagram, map, site plan, flow chart, Gantt chart)
Follow-up	Identify further information needed Comparison with similar cases Modelling Simulation
Closure	Write a letter of thanks to anyone who has provided an interview or given access

to pre-test is sometimes the cause of considerable problems with data acquisition or analysis later in the process. Emphasis is often placed on the need to pre-test interviews and questionnaires, but all methods should, wherever feasible, be based on some means of trialling.

Researcher question
Case Study

Do you intend to use a case study approach in your primary data acquisition?

Yes. I intend to use a case study approach and I am clear on the requirements of the process in order to use this method to an appropriate standard.
I understand that I may also need to consider ethical (and health and
safety) requirements.

No. The nature of my work does not lend itself to a case study approach.
I realise that if my topic or aims change, I may need to revisit this section.

Questionnaire and interview

A pilot study provides a trial run for research using questionnaires or interviews,
which involves testing the wording of the questions, identifying ambiguous
questions, trying out the technique to collect the data, measuring the effectiveness of the standard invitation to respondents, etc. Problems can be removed at
an early stage and content improved. Questions can be checked for clarity and
any issues with the nature of the questions addressed; also, areas which should be
added to the questionnaire or interview schedule can be identified. The mock
data from pre-tests can also be used to test the methods by which the real data
will be analysed. A pilot study requires an investment in time but will improve
response rates and the quality of the data.

Laboratory work

Pre-testing for laboratory work usually involves running the proposed method
at a much-reduced scale. Feasibility and timings are important as well as data
analysis. Attention should be given to any problems in the procedure which need
to be addressed, and any significant issues may warrant a second run prior to the
main data collection.

Fieldwork

The pre-data collection stage of work in the field varies considerably. This may
involve visiting a similar site to test collection methods, scoping the actual site
to allow concentration on specific areas or aspects, or visiting a site upon which
similar research has been undertaken in an attempt to unpick the data collection
processes used. It is recommended that before any fieldwork, the researcher is
fully acquainted with the methods to be used, the nature and extent of the site
and the underlying literature.

Supervisor guidance on primary data collection

Guidance from Supervisor A

The selection of primary data collection methods must be made with clear reference to your research objectives.

Clear consideration needs to be given to what you are investigating and whether the data you are going to collect needs to be quantitative and/or qualitative in nature. If your research is laboratory based, it is more than likely that the data you collect will be mainly quantitative (numeric) in nature, and the methods that you deploy to gather the data will be specified by European or British Standards or other such standard procedures.

If your research is concerned with developing theory, or if you are investigating workplace or management practices or behaviours, attitudes, opinions or relationships then it is possible that the data you collect will be mainly or entirely qualitative in nature.

A range of data collection methods may be available to the researcher. It is imperative that you consider: the appropriateness of each possible method with regard to the research objectives; the practical implications of each method; and the ethical implications of each method. Ensure that the methods you select are implemented with suitable rigour and with regard to validity and reliability in the process.

Primary data collection methods that are commonly utilised within dissertation research include: laboratory experiments, questionnaires, interviews, structured observation, participant observation, action research and case study investigation. Primary data case studies often combine a number of data collection methods to investigate a defined event, process, project or setting.

Once data collection methods have been selected, it is of critical importance that you give full consideration to the effective *management* of primary data collection. For laboratory experiments, the procedures, equipment and materials must be fully identified and approved by your supervisor before commencing data collection. If you are undertaking non-laboratory-based research investigations, you should have your research objectives and data collection methods approved by your supervisor. A number of universities formalise this process in the form of a 'research proposal'.

You should be able to explain why you are including each question within a questionnaire. Does each question support a research objective or does it help to classify the research participants? If not, why are you asking it? How are you going to promote a high response rate?

If you are gathering primary data using interviews, have you considered who you are going to interview and why? What research objectives are going to be addressed by the interview? Is the interview going to be conducted face-to-face or via another mechanism? When is the interview going to be conducted? In your interview, are you going to present the data gathered from other primary research methods for comment and response? How are you going to record the data that you gather at the interview?

If you are gathering primary data using observation of an event or setting, when is this going to take place? Have all observed participants provided consent? How are you collecting data? Are the data you are going to collect concerned with measuring the frequency or duration of events? Or is your data going to be concerned with describing interactions and transactions?

Whatever research methods you select to gather primary data, ensure that you can clearly answer the following questions (derived from a Rudyard Kipling poem) before you begin: What are you are going to do? Why you are going to do it? When you are going to do it? How you are going to do it? Where you are are going to do it? Finally, who is going to be involved?

Guidance from Supervisor B

I supervise students doing work both inside and outside the laboratory. For the laboratory work, it's relatively easy to help a student refine their initial ideas into something which will allow them to collect enough data to make a viable research project.

On the other hand, I do often get students who need a bit of help and guidance, so I usually recommend that they start with the literature, find out what has been done and see what ideas they can collect about what they might want to test.

The main issues are students trying to be either much too ambitious or, conversely, much too limited. I'd say that students need to get an early idea about what might be involved and what resources they might need. Sometimes students need to liaise directly with technical staff and perhaps even have small pieces of equipment built up for them. Obviously there are time issues with this, particularly since there could be several other students doing research all crying out for technical time. You might also need to remember that materials or equipment might have to be bought and that sometimes this might mean purchase orders having to be raised, signed off and more delays. You should just be aware that even with a practical project, there are elements of negotiation and resource control. It is important to realise that time will be a major factor and to start early so that you have every opportunity. Even with laboratory work, it's helpful to consider a contingency plan for if things don't work out as you had planned.

The main issue with delivering a laboratory-based project is making sure that enough time is allowed up front to adequately test all the specimens, leaving sufficient time to repeat any parts of the test where there is any doubt or to check your results by repetition. So if you are doing a laboratory-based piece of work, I would advise that you concentrate on the initial phases and make sure that these are as robust as you can make them. This investment will pay back later in the project.

If you are considering a field-based project, the health and safety side could make you decide that staying at home is the safer option! Really, it is all about considering what could go wrong and then trying to minimise risks. There is usually help and support available and getting data out 'in the real world' can really help to consolidate a lot of the theoretical aspects covered in university. If I was to pass on sound advice about doing anything fieldwork based in the UK, it would involve lots of practical advice. Make sure you have a warm, waterproof coat if you are going to be outside as it will rain. Make sure you have sunscreen, particularly if working near water. Ensure that you use a notebook which won't dissolve in the rain (Chartwell survey books are a good choice) and write with a pencil. Don't rely entirely on technology as it always fails when you need it. Pack spares (spare camera, spare batteries, extra power packs) of data collection equipment. Take loads of notes or you will forget things. Make sure you do lots of pre-planning and ask for advice early. I enjoy going to collect data in the field; it is challenging, but that just adds to the excitement I guess!

Guidance from Supervisor C

You need to get from point A to point B by a certain time, visiting certain points on the way.

There are a number of different routes which you can take – you can consider these to be the principle foci of your work or the key aims and objectives that you must meet.

There are a number of different strategies which you can use (train, bus, car, cycle, walk), some of which may be seen as more appropriate than others and some of which may not work very well at all (pogo stick). Time-wise you have a deadline to be there – yes, you can speed ahead and get there early, but there is a good chance that you will have skimmed through your destinations and not got enough out of them to really prove that you have visited them properly. Alternatively, you can spend so much time on the first destination that you then have to rush through the rest of them. You could leave it until the last minute, giving just enough time to go through each destination in sufficient detail. But the slightest problem (roadworks) will throw you off time, or you will have to work so fast that you undo all your good work by shoddy presentation or even plagiarism (the equivalent of getting caught speeding).

Just as a successful journey usually starts with a map, so careful planning of your research will greatly increase the likelihood of a positive outcome.

Logistically, you should have a clear idea not just about what data you want, but also how this will be collected and analysed. The lack of foresight from students regarding data handling and analysis has been the downfall of many an otherwise strong piece of work.

Student reflections on primary data collection

'I had a nasty shock when I realised I had sent out my questionnaire before I knew how to analyse the data. Had I known then what I learnt later about stats, I'd have designed my questionnaire very differently.'

'If you do a placement, a year out, shadowing or internships, you should look for possible case studies.'

'The work I did was practical and based outdoors. I lived in a youth hostel for about seven weeks, and that worked out really well for me. It did cost more than living on a campsite, but I knew the warden was keeping an eye on me and making sure I got back safe from the field.'

'I couldn't see why I needed to do an ethical review. I knew I wasn't going to do any harm with my research. Then my supervisor explained to me how my proposed research (talking to kids about the redesign of their school) might look to a parent! I was horrified that anyone could think that I had any other motives for talking to kids! Anyway, I just got the teachers to ask the questions instead and so I got my data without even going to the school. Result!'

'Doing laboratory work was really easy once the initial planning and organising was sorted out. I hadn't realised that I would have to try and do so much to get my work done, but once it was all moving, the process ran really smoothly.'

'The interviews were a really good opportunity to not only make contacts for the rest of your career, but give a lot of good and valuable information for your studies.'

'Make sure if you are going to do questionnaires that you don't presume people will answer your survey because they most probably won't, even if they imply they will.'

'I was surprised at how much professionals from companies were willing to help if you ask around nicely.'

'Setting up an online questionnaire was a very good idea. It allowed greater access to respondents. I used a paid site that allowed me to download the data into a spreadsheet. This saved me so much time!'

'You don't have to pay online charges to mail a questionnaire. I set up my questionnaire on Word and emailed this out.'

'With the interviews, I would suggest that you pre-book very early and make sure you do a lot of pre-planning to give yourself enough time, trying

> to see what you want from the data and what possible arguments would
> help the outcome of your dissertation so that you can work your question-
> naires/interviews around this.'
>
> 'I identified key players who were relevant in my subject area and tar-
> geted them for interview and questionnaire instead of questioning random
> people; this way, my candidates felt they had something special to offer.'

Summary

The chapter has provided an overview of the practical considerations required
for primary data collection. These include assessing and controlling any ethical
and health and safety considerations which may pose a risk for participants or the
researcher. An overview of good practice related to different primary data collection methods has been presented. The benefits of undertaking carefully planned
and rigorous research were outlined with suggestions to help the student manage
these processes.

Suggested further reading

BELL, J. (2010) *Doing Your Research Project: A guide for first-time researchers in education,
health and social science.* 5th ed. Maidenhead: McGraw-Hill Open University Press.

BIS (2007) *Rigour, Respect, Responsibility: A universal ethical code for scientists.* [Online].
Department for Business, Innovation & Skills and Government Office for Science (12
September). Available from: https://www.gov.uk/government/publications/universal-
ethical-code-for-scientists (accessed 1 August, 2015).

BOURQUE, L. B. & FIELDER, E. P. (1995) *How to Conduct Self-administered Mail Surveys.*
Thousand Oaks CA: Sage Publications. ISBN 0-8039-7388-8

CRESWELL, J. W. (2012) *Educational Research: Planning, conducting, and evaluating quantitative and qualitative research.* 4th ed. Boston MA: Pearson. ISBN 978-0-13-261394-1

DAWSON, C. (2009) *Introduction to Research Methods: A practical guide for anyone undertaking a research project.* 4th ed. Glasgow: Bell & Bain Ltd. ISBN 978-1-84528-367-4

DENSCOMBE, M. (2010) *The Good Research Guide for Small-scale Social Research
Projects.* 4th ed. Maidenhead: Open University Press/McGraw Hill. ISBN
978-0-335-24138-5

ESF (2011) European Code of Conduct for Research Integrity. [Online]. 4 March,
2011. Available from: www.esf.org/fileadmin/Public_documents/Publications/Code_
Conduct_ResearchIntegrity.pdf (accessed 1 August, 2015).

FARRELL, P. (2011) *Writing a Built Environment Dissertation: Practical guidance and examples.* Chichester: Wiley-Blackwell. ISBN 978-1-4051-9851-6

FELLOWS, R. & LIU. A. (2008) *Research Methods for Construction.* 3rd ed. Chichester:
Wiley-Blackwell. ISBN 978-1-4051-7790-0

GILL, J., JOHNSON, P. & CLARK, M. (2010) *Research Methods for Managers.* 4th ed. Los
Angeles; London: SAGE. ISBN 978-1-84787-094-0

MARDER, M. P. (2011) *Research Methods for Science.* Cambridge: Cambridge University
Press. ISBN 978-0-521-14584-8

RCUK (2013) The Research Councils UK Code of Good Research Conduct. [Online]. Updated July 2015. Available from: www.rcuk.ac.uk/Publications/researchers/grc/ (accessed 1 August, 2015).

ROBSON, C. (2011) *Real World Research*. 3rd ed. Hoboken, NJ; Chichester: Wiley. ISBN 978-1-405-18240-9

SMITH, K. (2009) *Doing Your Undergraduate Social Science Dissertation*. London: Routledge. ISBN 978-0415467490

SWETNAM, D. (2001) *Writing Your Dissertation: How to plan, prepare and present successful work*. 3rd ed. Oxford: How To Books Ltd. ISBN 1-85703-662-X

UNIVERSITIES UK (2012) The Concordat to Support Research Integrity. [Online]. Available from: www.universitiesuk.ac.uk/highereducation/Documents/2012/The ConcordatToSupportResearchIntegrity.pdf (accessed 1 August, 2015).

WALLIMAN, N. (2011) *Research Methods: The basics*. Abingdon: Routledge. ISBN 978-0-415-48994-2

WCRI (2010) Singapore Statement on Research Integrity. [Online]. *World Conference on Research Integrity*. Posted 22 September. Available from: www.singaporestatement.org/ (accessed 1 August, 2015).

WMA (2013) Declaration of Helsinki. [Online]. Adopted by the 18th WMA General Assembly, Helsinki, Finland, June 1964 and amended by the 64th WMA General Assembly, Fortaleza, Brazil, October 2013. © 2015 World Medical Association, Inc. Available from: www.wma.net/en/30publications/10policies/b3/ (accessed 1 September, 2015).

8 Data analysis and presentation

Introduction

This chapter presents a range of approaches to data analysis and, in so doing, discusses and outlines:

- qualitative data analysis and its use of coding and categorisation in the undertaking of ethnographic, phenomenological and grounded theory studies;
- using quantitative techniques to illustrate and present qualitative data;
- quantitative data analysis techniques – approaches to the numerical and statistical analysis of data;
- virtual supervisor guidance regarding data analysis;
- some real-life student reflections regarding analysis and presentation of data;
- further reading.

The analysis of data is a core component of the dissertation process. Appropriately selected and rigorously applied methods of data analysis are necessary for all dissertation studies, regardless of whether your dissertation is purely a desktop study of existing data, a study of 'big data', or one that also involves the collection of primary data to inform a scientific or interpretive study.

A range of data analysis methods are available to researchers. This chapter concisely considers and discusses qualitative and quantitative analytical methods. Your choice of analytical methods for your dissertation should be informed by a number of related factors. These include: the aim of your dissertation; whether your research sits within the scientific or interpretive domain; and the intended meaning of your research results.

Data analysis (whether the data are of a quantitative or a qualitative nature) may be approached inductively or deductively. In general terms, analysis that is inductive starts from the data and develops concepts and theories based on the data (see 'grounded theory' in Table 8.1). The aim is to produce meaningful insights and rich descriptions that are specific to the research context. Analysis that is deductive is based on testing existing theories/hypotheses. Taking this approach, the purpose is to discover or construct universal truths or results that are generalisable outside the context of the research study (see 'content analysis'

Table 8.1 Two techniques commonly used in undergraduate dissertations

Technique	Process
Grounded theory	Grounded theory is inductive; data collection, analysis and theorising are part of an iterative process
	The collected text (notes and/or transcripts) are read
	The most important issues from the participant's point of view are identified (open coding)
	Themes are noted as they emerge and are placed into categories
	Categories are used to create theoretical modes which are compared to the literature
Content analysis	Content analysis can be deductive
	Variables have been isolated from the literature
	The collected text is then compared to these variables
	Themes may be quantified and the results expressed statistically
	The outcomes can be compared to those of other measures

in Table 8.1). Although qualitative analysis techniques are often associated with an inductive approach and quantitative with a deductive approach, this distinction is not firmly held. For example, qualitative research can draw on theories and concepts that already exist in the literature.

Analysing qualitative data

'Qualitative data analysis' is a term that describes a number of analytical methods or practices. These methods and practices are concerned with the systematic interpretation and organisation of rich qualitative data.

Qualitative data analysis and the intended meaning of research results

The researcher's approach to both qualitative data collection and quantitative data analysis is informed by the intended meaning of the research results. Broadly, qualitative analysis is based on the inductive approach. If the intention is to produce a description of a 'lived experience' of something then a *phenomenological* approach is adopted. An *ethnographic* approach to data collection and analysis is adopted when the intention is to produce a rich detailed description of a 'social environment' and its 'culture'. Alternatively, if the intention is to produce 'theory' that emerges from a study of a social environment then a *grounded theory* approach may be adopted.

Phenomenological studies

Phenomenological investigations employ data collection methods that seek to capture descriptions of the lived experience of participants in relation to a defined phenomenon. That phenomenon might be the lifeworld of a tower crane operator, the experience of surveying tunnels, communication with project supply chains, quality control procedures on construction sites or delivery of site inductions on major construction projects.

Interviews are a common mainstay of data collection for phenomenological studies. Descriptions written by participants of the phenomenon, or other expressions of the lived experience, possibly poems or drawings, can also provide meaningful and useful data sources for the production of a phenomenological narrative.

The categorisation and coding of documented data enables the researcher to identify key themes that emerge from the data. Through the process of analysing occurrences of the emergent themes, the researcher strives to present structured, rich descriptions that bring insight and meaning to the phenomena of study. These rich descriptions are illustrated and supported by examples and quotations drawn directly from the primary data.

Ethnographic studies

Ethnographic studies involve the study of people and culture within their natural setting and seek to provide a rich detailed description of the social environment. Data analysis involves the use of coding and the development of themes to assist in delivering rich, meaningful description of the investigated social environment. The presentation of an ethnographic study should juxtapose rich description and observations alongside quotations and field note excerpts drawn directly from the social environment of the study.

Grounded theory

With a grounded theory approach, analysis is undertaken with the purpose of developing 'theory' that is emergent from the research study. Barney Glaser and Anslem Strauss are the originators of grounded theory, having developed this approach to data and theory development in the 1960s. Data collected from a phenomenon of study are categorised and coded. The researcher revisits the data time and time again to refine categorisation and re-code. Further data are collected on a number of occasions to verify or further refine the emergent theory. The analytical approach to grounded theory development can be described as consisting of two methods, these being:

- *Constant comparative method* – This requires categories and concepts emergent from newly collected data to be constantly compared with that of the data already collected. Theory is proposed as categories and concepts emerge and are refined.
- *Theoretical sampling method* – The theoretical sampling method uses the data collected and analysed initially to guide and inform the focus of the researcher's further data collection and analytic abstraction. Key themes that emerge from the data provide the researcher with a focus as to what data to collect next and where to find it. Here, the coding, categorisation and analysis of

data serve to guide and inform the search for further data from the field of study and the refined, focused generation of theory.

Coding and categorisation

The first stage in the qualitative analysis process is to become familiar with the material by reading and rereading the transcripts and/or reviewing observation data and making notes of the key ideas, patterns and themes as they emerge. It is not necessary to wait until all the data are collected; collection and analysis are best done in parallel, each informing the other.

Coding and categorisation are analytical techniques that are used to organise and arrange qualitative data that have been collected and documented in observation notes, fieldwork notes and interview transcripts. The technique requires that key themes or categories are identified from the text, and the text is then coded according to these categories. This requires the researcher to be immersed in the data in order to identify categories and themes. Subtle differences need to be demonstrated. For example, a staff restructuring might be coded as 'change', but it may be more meaningful to note specific concerns that come up as part of restructuring, regarding relocation ('change place'), pay alterations ('change pay') or changes in social grouping ('change people'). In addition, these changes could be coded as positive or negative from the interviewee's perspective. This allows a large volume of text to be reduced to a set of data which is far easier to compare and contrast. Coding is then carried out by systematically indexing the data using the categories identified.

Qualitative data analysis software has been developed to assist with the process of coding, and this is now an option commonly used by qualitative researchers. An ever-increasing number of software packages are available. These include ATLAS.ti, MAXQDA and NVivo. It should be noted, though, that qualitative data analysis software does not undertake all of the requisite analysis on behalf of the researcher. The software helps and assists with sorting, structuring, managing and presenting the data. The researcher remains the filter for the analytical process that must be applied to data organised and managed using the software. The use of software does bring potential for great time saving. In addition, it can produce graphical illustrations and provide enhanced audit-ability of the process.

Presenting qualitative data

Qualitative data analysis supports the development of rich meaning and description. In order to promote meaning and enable richness in description, great care needs be taken to ensure quotes and descriptions taken from the qualitative raw data are presented with full reference to emerged themes and categories. Failure to do so can result in the dissertation presenting a list of disparate information rather than a coherent narrative.

Tabulation of data enables the presentation of data in a structured way. Through the considered use of tables, qualitative data can be clearly arranged,

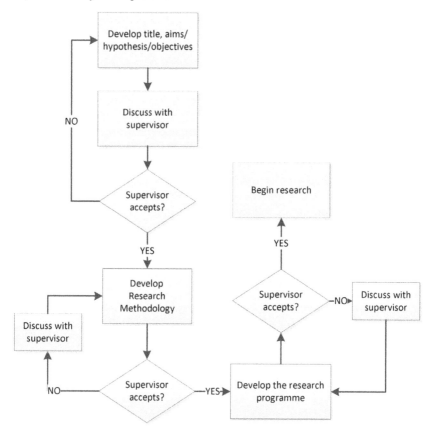

Figure 8.1 Flow chart illustrating aspects of the dissertation research process

themed and stratified. Flow charts are also commonly utilised to illustrate, map and communicate processes, as illustrated in Figure 8.1.

Techniques for presenting qualitative data

Whilst tables and flow charts are commonly used to organise and arrange the presentation of qualitative data, there is a variety of visualisations that can provide a useful way of summarising this often complex information. They can assist the reader to easily view evidence and information which supports the analysis. A good visualisation can reduce the need for excessive descriptive text. Even relatively simple visualisation techniques such as use of shading or colour (Figure 8.2) can help to improve the clarity of information presented within the thesis.

There may be occasions when quantitative data presentation techniques assist in providing richness to the understanding of a qualitative phenomenological,

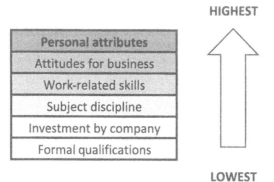

Figure 8.2 Final ranking of importance of candidates' skills and attributes during employee selection

ethnographic or grounded theory investigation. Quantitative-based visuals such as Venn diagrams and Wordles may also serve to assist in the presentation of meaning.

Venn diagrams can be used to illustrate and communicate the relationships between groups or sets of data. Such diagrams can prove to be a very useful tool in presenting relationships between themes, attitudes, opinions and perceptions. Figure 8.3 provides an example of a Venn diagram that illustrates interviewees' perceptions of where key skills and knowledge are assessed within the interview process.

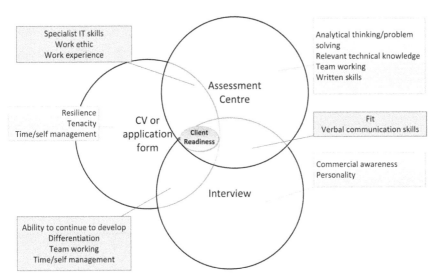

Figure 8.3 Venn diagram illustrating interviewees' perceptions of where skills and attributes are assessed within an assessment process

Word clouds (Wordles) can also be a useful illustrative tool for qualitative data. They show the most frequently occurring words within a body of text. The greater the occurrence of a word, the larger in size the word is within a Wordle illustration. As such, Wordles can visually illustrate the relative predominance of a research response or a theme. Whilst Wordles do not show the relationship between words, they can be a useful technique when reviewing data for further analysis or in simply presenting the comparative frequency or predominance of responses. An example of a Wordle is presented in Figure 8.4. This was developed and generated with the use of online Wordle tools at http://worditout.com/.

When thinking about using and incorporating visual illustrations into the dissertation to support the presentation and analysis of data, the following considerations should be made:

- A visualisation should be carefully chosen and constructed so as to contribute meaningfully to achieving the dissertation aims and objectives or to examining the hypothesis.
- A graph or visualisation should add clarity to the presentation of data, analysis or findings.
- Visualisations should help the reader to better understand the research data, analysis and findings.
- Visualisations should be suitably sized, clearly labelled and appropriately located within the dissertation.

Figure 8.4 A word cloud illustrating the frequency of words used by students to describe quantitative data analysis

- In some instances, text or a table can be better suited and more appropriate to the communication of meaning than a graph or other visualisation.

The visual presentation of data should not be confused with analysis. It should be understood that it is possible to include a variety of visual aids within a dissertation which add little or nothing to the fulfilment of the research objectives or the reader's understanding.

Analysing quantitative data

The goal of data analysis is to make the strongest possible case from your data. Sample size and variability of sample may influence your interpretation, as might your brain's natural attempts to find patterns (even when they are not actually there). Using statistics helps to overcome these potential biases.

Structured quantitative data can be analysed using:

- *Descriptive statistics* – These provide a way to summarise the data. They do not permit conclusions to be made beyond the data analysed. They allow data to be summarised in a meaningful way, which may uncover patterns. They include measures of central tendency (mean, median, mode) and measures of dispersion (range, inter-quartile range, standard deviation, variance, coefficient of variation).
- *Inferential statistics* – These allow inferences to be made about a population based on the data collected from a sample of that population. They also show how much confidence there is in a sample.

Organising data using MS Excel

In the section that follows, instructions are included for producing descriptive statistics or charts using Excel. Excel is recommended for students unfamiliar with statistics as learning to use it is relatively easy, and it has an array of easily accessible help functions within the program and accessible on the Internet.

When inputting data, it is recommended that a single master worksheet is used for all raw data. No analysis should be done on this sheet; instead, a copy of the data should be taken and moved into a separate workbook.

Questionnaire data sets should be arranged so that the participant information is on an individual row. It is suggested that any statistical summaries (mean, mode, standard deviation) are done along the top rows. This saves time in scrolling down large data sets to the bottom of a sheet. There are a number of functions which will be helpful and that you should be familiar with:

- inserting a graph;
- use of the Sort function;
- use of the Filter function;

- use of Conditional formatting;
- how to access the Help function.

Quantitative data description

Descriptive statistics most commonly used by dissertation students are summarised in Table 8.2 and these are among those covered in more detail below. Refer to Table 8.3 for guidance on which techniques are suitable for the types of data you want to analyse as part of the dissertation.

Mean

The mean is the arithmetic average and the simplest descriptive statistic. To calculate it, add up all the points of data and divide by the number of points of data. For interval or ratio data, the mean and standard deviation are a good place to start your description. The average value is sensitive to the occurrence of a few incidences of very high or very low values.

Using Excel

=AVERAGE(first cell of column:last cell of column)

This is illustrated in Figure 8.5, where the formula used to calculate the mean is =AVERAGE(B5:B15).

Median

The median is the value that occurs at the halfway point if all the values were ranked in order. Where there are an even number of values, the median is the average of the two values either side of the halfway point. The median value is a good way of measuring central tendency where there are a few very high or very low values (as in Figure 8.9).

Table 8.2 Summary of basic descriptive statistics

Mean	The arithmetic average
Median	The centre point of the distribution where half of the values are higher and half lower
Mode	The most frequently occurring value
Range	The difference between the highest and lowest value
Inter-quartile range	The difference between the upper and lower quartiles
Standard deviation	The square root of the variance
Coefficient of variation	Interprets the relative magnitude of the standard deviation by dividing it by the mean; expressed as a percentage
Variance	A measure of spread, calculated as the mean of the squared differences of the observations from their mean

Table 8.3 Techniques for describing different types of data

Where the distribution is important

Type of data	*Suggested techniques*
	Graphs:
Nominal	Pie chart
Nominal/ordinal	Column chart or bar chart
Ratio/interval	Histogram

Where the central tendency is important

Type of data	*Suggested techniques*
	Descriptive statistics:
Nominal	Mode
Ordinal	Median
Ratio and interval	Mean

Where the spread of data is important

Type of data	*Suggested techniques*
	Descriptive statistics:
Ordinal	Range
	Inter-quartile range
Ratio/interval	Range
	Inter-quartile range
	Variance
	Standard deviation
	Coefficient of variation
	Graphs:
Ratio/interval	Box and whisker

Using Excel

=MEDIAN(first cell of column:last cell of column)

Mode

The mode is used with nominal data as a measure of central tendancy. It is the most common value that appears in the data.

Using Excel

=MODE(first cell of column:last cell of column)

Standard deviation

Standard deviation (SD) gives a measure of the variability of the data. There are two equations to calculate SD for the whole population (N) or the sample

	A	B	C	D
1			Formula	
2	Average		3 =AVERAGE(B5:B15)	
3				
4		Data set		
5		4		
6		4		
7		5		
8		4		
9		2		
10		2		
11		3		
12		1		
13		5		
14		1		
15		2		
16				
17				
18				

Figure 8.5 Using MS Excel to calculate the mean

population (N-1), and it is recommended that you use the latter calculation if you are not certain. The smaller the standard deviation, the more representative the mean is to the population. A standard deviation of 0 will be returned where all the values are the same.

Using Excel

=STDEV(first cell of column:last cell of column)

For a data set which is distributed normally, 68 per cent of the data will occur in the range which lies 1 SD above and 1 SD below the mean. This will always be the case for a normal distribution, regardless of whether the values are clustered around the mean or dispersed more widely (Figure 8.6).

Standard error of mean

Also known as SEM, the standard error of mean gives a range based on the variability of the data in your sample in which the true mean of the underlying population will occur. To calculate it, divide the SD by the square root of the

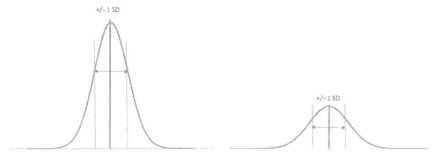

Figure 8.6 Frequency distribution for a clustered data set (left) and a dispersed data set (right)

sample size. Interpretation of the SEM on its own is difficult, but it can be used to calculate the confidence interval (CI; see below).

Using Excel

Calculate standard deviation
Calculate sample size n using the formula = COUNT(first cell of column:last cell of column)
= Cell1/(SQRT(Cell2))
where Cell1 is the cell containing SD and Cell2 is the cell containing n

Figure 8.7 shows the effect of larger data sets on the SEM and the CI. The average, mean and standard deviation in this case are unaffected by the increased number of data points. CI and SEM reduce as the number of points increases because it is possible to become more certain about the nature of the underlying population.

Confidence interval

The calculated mean from your sample is not necessarily the same as the population mean. The confidence interval considers sample size and variability of the sample to give a range around the mean with a 95 per cent confidence interval (95 per cent is used most commonly, but this can vary). If the population is Gaussian (conforms to the bell-shaped curve) then you are stating that in 95 out of 100 times, the true population range will be likely to occur within this range of values (Figure 8.8).

Using Excel

=CONFIDENCE(significance level,standard deviation,size)
significance level is usually taken to be 0.05 (confidence level of 95 per cent)
=CONFIDENCE(0.05,Cell1,Cell2)
where Cell1 is the cell containing SD and Cell2 the cell containing n

	A	B	C	D	E	F	G	H	I	J	K	L	M	N	O	P	Q	R
1		Data set 1		Data set 2														
2	Average	4.0		4.0														
3	Median	4.0		4.0														
4	**Sample size (n)**	**12**		**180**														
5	Standard deviation	1.08		1.08														
6	**SEM**	**0.31**		**0.08**														
7	**CI**	**0.61**		**0.16**														
8																		
9		Data set 1		Data set 2														
10		2			2	2	2	2	2	2	2	2	2	2	2	2	2	2
11		4			4	4	4	4	4	4	4	4	4	4	4	4	4	4
12		5			5	5	5	5	5	5	5	5	5	5	5	5	5	5
13		3			3	3	3	3	3	3	3	3	3	3	3	3	3	3
14		6			6	6	6	6	6	6	6	6	6	6	6	6	6	6
15		5			5	5	5	5	5	5	5	5	5	5	5	5	5	5
16		3			3	3	3	3	3	3	3	3	3	3	3	3	3	3
17		4			4	4	4	4	4	4	4	4	4	4	4	4	4	4
18		4			4	4	4	4	4	4	4	4	4	4	4	4	4	4
19		4			4	4	4	4	4	4	4	4	4	4	4	4	4	4
20		3			3	3	3	3	3	3	3	3	3	3	3	3	3	3
21		5			5	5	5	5	5	5	5	5	5	5	5	5	5	5
22																		
23																		

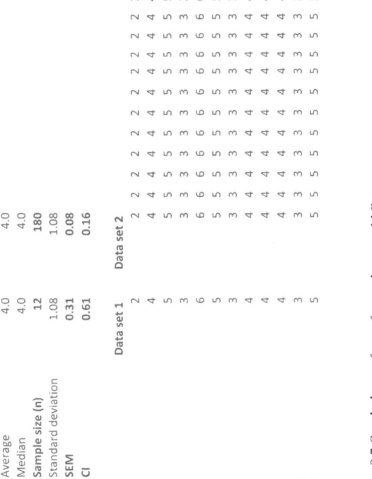

Figure 8.7 Standard error of mean for two data sets of differing size

	A	B	C	D
1			Formulae	
2	Average	4.4 = AVERAGE (B10:B21)		
3	Median	4.5 = MEDIAN (B10:B21)		
4	Sample size (n)	12 = COUNT (B10:B21)		
5	Standard deviation	1.55 = STDEVP (B10:B21)		
6	SEM	0.45 = B5/(SQRT(B4))		
7	CI	0.88 = CONFIDENCE(0.05,B5,B4)		
8				
9		**Data set**		
10		2		
11		4		
12		6		
13		3		
14		6		
15		6		
16		6		
17		3		
18		4		
19		2		
20		6		
21		5		
22				
23				

Figure 8.8 MS Excel formulae for common descriptive statistics

Quartile and inter-quartile range

The 'range' of values includes all data even when there are only a few occurrences of very high or very low values (as shown in Figure 8.9). In this case, the minimum and maximum values are not typical of the majority of the data collected. The inter-quartile range gives a more representative view of the data set by removing unusually high or low values; it does this by measuring the dispersal of the central 50 per cent of values based upon the median (the central) value. The quartiles split the data into four – the lowest 25 per cent, the highest 25 per cent, and then 25 per cent above and below the median.

Using Excel

=QUARTILE(first cell of column:last cell of column,1) – calculates the 1st quartile
=QUARTILE(first cell of column:last cell of column,3) – calculates the 3rd quartile

In the example in Figure 8.9, the lower quartile is values of 46.75 or below, the upper quartile is values of 65 or above, and the inter-quartile range (where 50 per cent of the values are located) is between scores of 46.75 and 65.

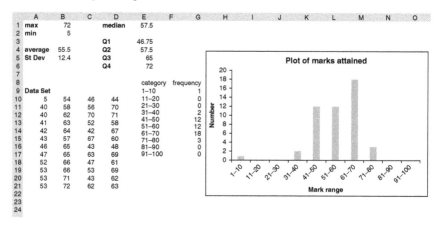

Figure 8.9 An MS Excel sheet showing average, range, inter-quartile range and standard deviation

Other useful excel formulae:

=SUM(first cell of column:last cell of column) – adds all values together
=MAX(first cell of column:last cell of column) – returns maximum value
=MIN(first cell of column:last cell of column) – returns minimum value

Skew and kurtosis

Skew and kurtosis provide further information about the distribution of data. Skew refers to a distribution that is not symmetrical. A distribution can be positively or negatively skewed (Figure 8.10). Skew can be given a numerical value using the formula (mean – mode)/standard deviation. Kurtosis is the term applied to the shape of the distribution curve – this might be narrow and pointed, or broad and flat.

Presenting quantitative data

Visual aids such as graphs and charts are useful ways of presenting data, but these have to be interpreted and you should provide commentary to tell your reader

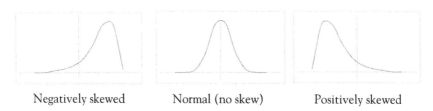

Negatively skewed   Normal (no skew)   Positively skewed

Figure 8.10 Symmetrical and skewed distributions

what the graph or chart does (or does not) show. The way in which you present your data should help to answer a question and communicate patterns, associations or trends. It should not distort the data and should be effective in conveying the message in an appropriate way to the audience. Graphs should all be shown in the main text of your report, and there should be a clear reason for inclusion.

Consideration of the data and the best forms of presentation should always be made. In some cases, words or a table will be able to communicate the message better than a graph or other visualisation. If there are only a limited number of data points (e.g. the relative percentages of two categories), a short sentence will suffice. For more complex data, use a table, graph or some other visual aid.

There are some conventions to follow when including graphs, tables and other visualisations in your dissertation. A caption (i.e. title) should always be provided for a table, graph, etc., and this should concisely and clearly reflect the content. Another convention that applies across the board is that illustrations should always be clearly referred to in the text so that the reader can understand why it has been included.

Creating tables

Tables must be easy to read and should not try to convey too much information. A table should be presented on a single page if possible. Points to consider when creating a table include:

- Include clear labelling.
- Avoid inclusion of too much text and/or text that is too small.
- Avoid lists of numbers with no label (e.g. where the numbers could be the mean or the standard deviation).
- For tables including numbers, do not use too many decimal places.
- The same formats (font, number of decimal places, etc.) should be used for all tables in the dissertation.

Using colour or shading on a table can increase clarity if used appropriately (see Figure 8.11).

A traffic light visual uses red, green and amber 'lights' and can be used to indicate performance against a benchmark (see Figure 8.12). For black and white documents, it is recommended that the colours are replaced with black/grey/white as colours do not always print clearly in greyscale.

Colour coding using Excel

1. Select the data
2. In the Format menu, select Conditional formatting
3. Click on Add a rule
4. Make a selection under Style (e.g. Icon sets as shown in Figure 8.12)
5. Select the required set

	Q1	Q2	Q3	Q4
R1	4	2	2	5
R2	3	5	2	2
R3	4	3	2	2
R4	1	4	5	2
R5	5	4	4	2
R6	1	2	3	2
R7	5	1	1	4
R8	4	1	4	2
R9	1	4	4	4
R10	2	1	2	4

Figure 8.11 Use of shading to enhance a table (created in MS Excel)

	Q1	Q2	Q3	Q4
R1	3	2	3	2
R2	1	2	1	4
R3	4	5	5	4
R4	1	2	3	3
R5	2	3	5	5
R6	2	4	2	3
R7	4	3	1	4
R8	4	4	2	2
R9	4	1	2	2
R10	2	4	1	3

Figure 8.12 Use of icons to create a traffic light diagram (created in MS Excel)

Creating charts

A chart is a way of showing data by graphical means. Charts can be used to show qualitative or quantitative data. They can take various forms, as outlined below; for example, graphs are a type of chart used predominantly for quantitative data.

The visual aid should be relevant to the points under discussion in the text. However, explanatory text should not simply repeat the message communicated in the graph; rather, there should be some degree of interpretation. Charts should be simple (limit the number of data sets) but include an appropriate level of detail. Rather than overcrowding a graph with data, it may be better to produce a larger number of graphs with simpler content.

Good practice in presentation of charts includes:

- Ensure all axes are labelled (on the graph or using a clear legend, with units of measurement given) and that labels may be read clearly.
- Axes start at 0 unless there is a good reason otherwise (e.g. all data have very high values).
- Selection of variables to be plotted should be considered (e.g. it is not usual to use dependent variables on both *x* and *y* axes).
- Where comparisons are to be drawn between graphs consider the scale on the *y* axis. Data can easily be distorted where scales are different.
- Avoid meaningless graphs (e.g. a pie chart showing that 50 per cent of respondents agreed and 50 per cent disagreed is unnecessary).
- Appropriate colours should be used for the version that will appear in your dissertation. If your dissertation will be reproduced in black and white, choose on-screen colours carefully so that the reader can still make sense of the graphic.
- Avoid borders or shadowing effects around the graph, unnecessary labels or lines, excessive text, fussy backgrounds and unnecessary 3D effects.

Generic process for creating charts using Excel:

1. Select the data
2. From the Insert menu, select Chart
3. Select chart type
4. Use chart tools to customise (e.g. labels, legend, colours)

Types of chart

The type of chart used can be decided with reference to the data and to the purpose of the analysis. The techniques that are appropriate for the different types of data are outlined in Table 8.3, and a range of different types of chart are outlined below.

PIE CHARTS

Pie charts are used to show the percentage contribution to the whole of categorical, ordinal or grouped ratio/interval data. They should be used with caution as column or bar charts are often better. The largest segment should be at 12 o'clock going clockwise (see Figure 8.13); however, plotting software will often start with the first category at 12 o'clock. Figure 8.13 shows the pie chart in 2D and 3D form. Note the use of 3D visualisation here can distort the perception of the data, making comparison of the values difficult.

Figure 8.13 Pie charts (created in MS Excel)

COLUMN/BAR CHARTS

Column charts or bar charts are used for displaying discrete data (nominal or ordinal). These charts are the same, but the former plots data vertically while the latter plots data horizontally. It is largely a matter of preference which is used, though the bar chart format is useful if the variables have long names or where a variable has many categories. Column/bar charts can show data for one or a small number of variables at a time

When you choose to include column/bar charts, there are various formats to choose from (see two basic formats in Figure 8.14). As illustration of how the same data (shown in Table 8.4) might be presented using different formats, see Figure 8.15, showing:

- *Clustered column chart* – This focuses on comparison of categories across different groupings.
- *Stacked column chart* – This shows overall differences as well as the relative compositions for each grouping.
- *Percentage bar chart* – This focuses on comparison of the proportions that make up the whole across different groupings.

Examples of stacked column charts are given in Figure 8.16 as follows:

- *Stacked 100% column* – Values for each variable are shown as a proportion of 100 per cent. This is used when the relative differences between the groups is to be compared.
- *100% stacked column with subcomponents* – This is used to give more detail for specific categories.
- *100% stacked column* – This may be used to show change over time or to compare relative compositions across groupings. A 100% stacked column (or bar) chart can be misleading where the underlying values of each class are different. Where possible, this should be labelled on the chart.
- *Stacked column chart* – As noted above in relation to Figure 8.14, this compares the overall differences between variables and shows the proportions that make up each of the variables. This is commonly used to illustrate change over time.

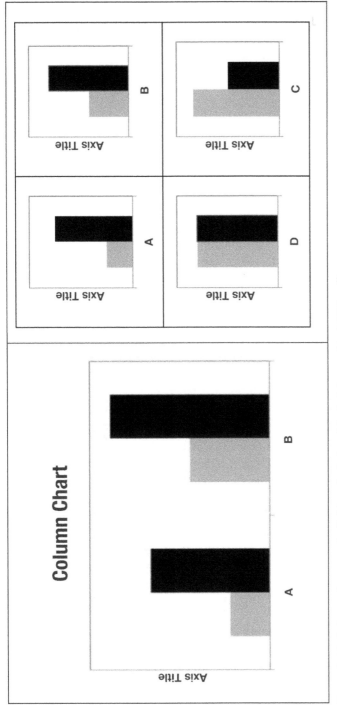

Figure 8.14 Column chart, single (left) and tabulated (right) (created in MS Excel)

Table 8.4 Raw data to be presented as a column/bar chart

	Group 1	Group 2	Group 3
Class A	5	2	1
Class B	3	1	0
Class C	10	5	1
Class D	9	4	2
Class E	8	3	2

It is most likely that only one or two of these would be presented to the reader in the dissertation, depending on the message that the student most wants to convey. It is advised that for any one series of charts, the chart format does not change so as to allow the reader to compare results.

A general rule is that ordinal data should be shown with categories in their logical order. However, when nominal data are displayed, columns/bars may be arranged in any order. Consider arranging bars/columns by size order unless there is a reason not to; for example, the data need to be presented chronologically or alphabetically (Figure 8.17). Bar/column charts may be used to show positive or negative values (or deviations from a set baseline value) as in Figure 8.18.

HISTOGRAM

A histogram is used to describe the distribution of variables measured at ratio or interval levels. Where there are large ranges of data, these are grouped into fewer bands so that the graph is easier to read. Figure 8.19 shows a bar histogram and a line histogram illustrating the shape of the distribution.

LINE CHART

Line charts are used to show trends or relationships between ordinal or ratio/ interval data. The independent variable is given on the *x* axis and the dependent variable (or variables) is on the *y* axis. Line charts are particularly useful when comparing changes to one or more variable over time. Trend lines can be added if appropriate (Figure 8.20). Although multiple dependent variables can be plotted, the graph can start to become confusing if there are more than four.

AREA CHART

An area chart, similar to a line chart, can show changes over time, but with an area chart, the area below the line is filled in. Examples of area charts are given in Figure 8.21:

- *100% stacked area chart* – This is commonly used to consider relative changes in composition over time.

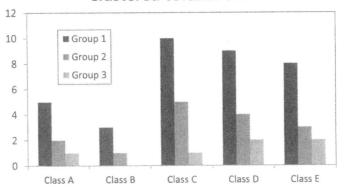

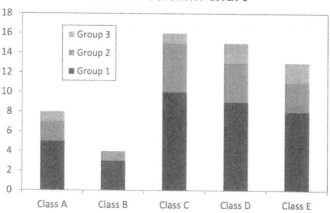

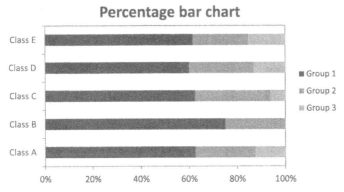

Figure 8.15 Clustered column chart, stacked column chart and percentage bar chart (created in MS Excel)

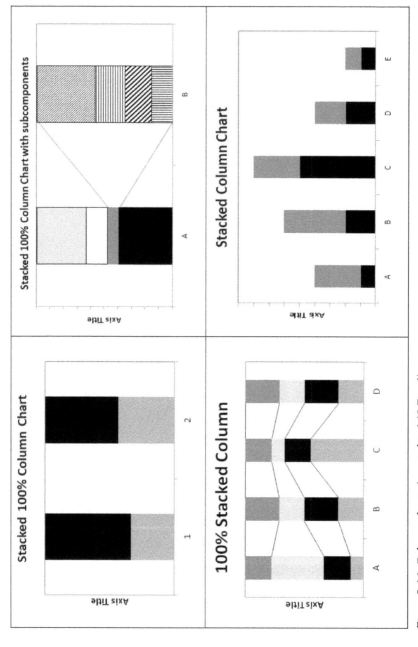

Figure 8.16 Column charts (created in MS Excel)

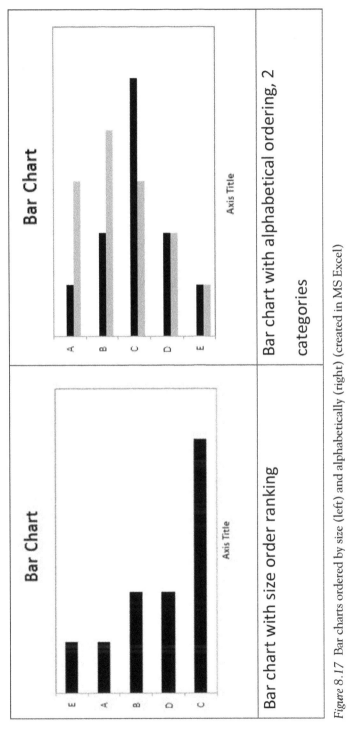

Figure 8.17 Bar charts ordered by size (left) and alphabetically (right) (created in MS Excel)

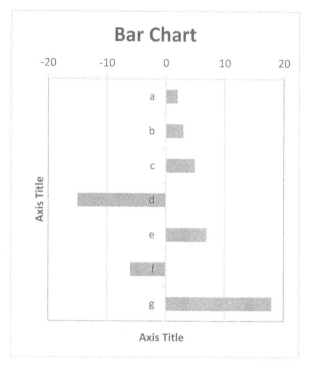

Figure 8.18 Bar chart with positive and negative values (created in MS Excel)

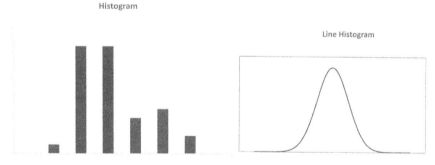

Figure 8.19 Example histograms (created in MS Excel)

• *Stacked area chart* – This is commonly used to consider absolute changes in composition over time. One set of data should be consistently smaller than the other for clarity.

RADAR CHART

Like a line chart, a radar chart plots variables against one another, but the latter has an axis for each category of an independent variable radiating from a centre

Line Chart

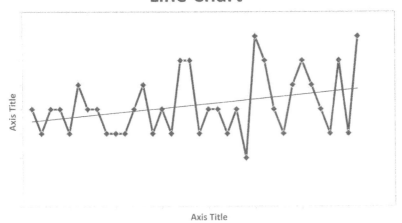

Line chart with trend line

Line Chart

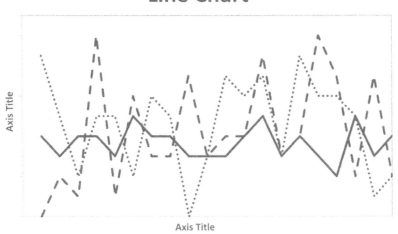

Line chart with multiple data sets

Figure 8.20 Line charts (created in MS Excel)

point. Data for the dependent variable are plotted on these axes and the points then joined together by lines. Radar charts can be difficult to read so are not often used. The example shown in Figure 8.22 shows occurrences for a dependent variable at different times over the course of 24 hours. Multiple dependent variables can be plotted to compare occurrences at the same time of day.

100% Stacked Area Chart

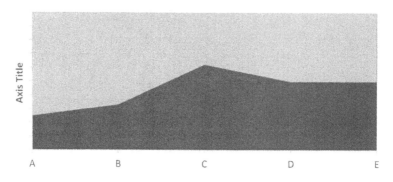

Stacked Area Chart

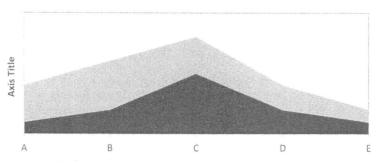

Figure 8.21 Stacked area charts (created in MS Excel)

Radar

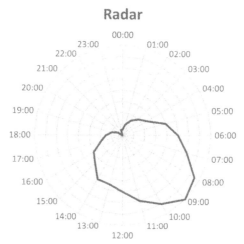

Figure 8.22 Radar chart (created in MS Excel)

VARIABILITY CHART

This may be used to compare the summary of descriptive or statistical outcomes for sets of data; for example, Figure 8.23 shows mean, high and low values. An expansion of this is the box and whisker plot where the median and inter-quartile range are shown with 'whiskers' representing the overall range.

SCATTER CHART

This is used to establish relationships between two variables where there are many points of data. Trend lines may be used (Figure 8.25).

A ternary plot is similar to a scatter plot and used where there are three variables and many points of data. Groupings of data may be distinguished within the ternary plot – an example is given in Figure 8.24, created using Tri-plot by Graham and Midgely (2000).

ROSE DIAGRAMS

A rose diagram displays directional data (degrees from north) and the frequency of occurrence in each class (Figure 8.26a). It is used commonly in Geography and Environmental Sciences, for example, in showing directions of strike in rock strata (Figure 8.26b).

Other methods

The nature and focus of the topic may be one where other methods of visualisation and presentation are appropriate. These may include: other 2D visualisations, such as maps or plans, specifications or guidance notes; or production of

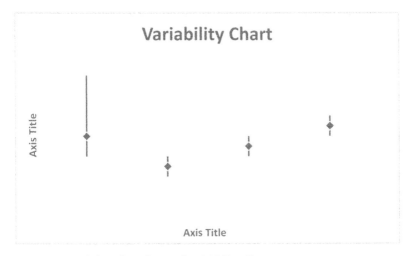

Figure 8.23 Variability chart (created in MS Excel)

Scatter Chart

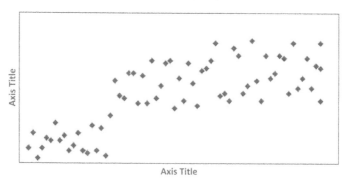

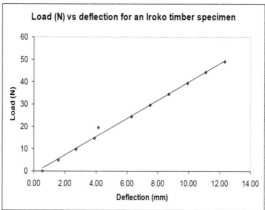

Figure 8.24 Scatter charts (created in MS Excel)

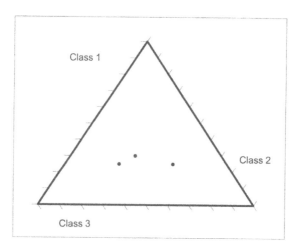

Figure 8.25 Ternary plot (created in MS Excel using Tri-plot by Graham and Midgely, 2000)

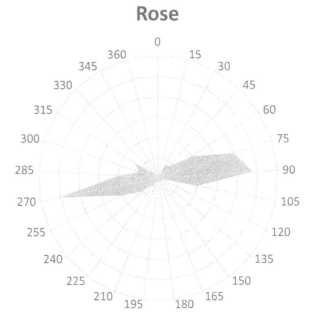

Figure 8.26a Rose diagram (created with MS Excel)

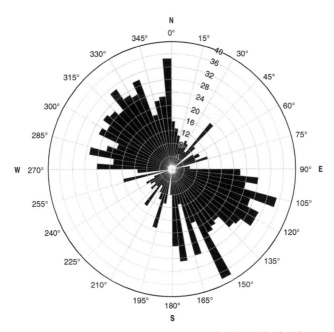

Figure 8.26b Dip directions (plotted using Georose by Yong Technology Inc., 2014)

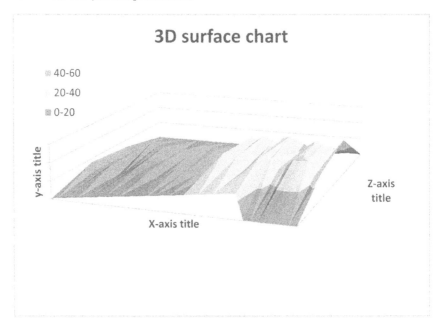

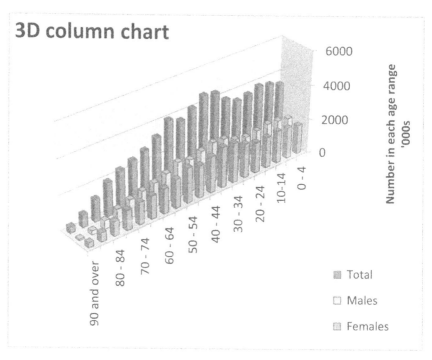

Figure 8.27 3D charts (created in MS Excel)

physical or virtual static 3D models, 3D fly-through models. As noted above, use of 3D visualisations can produce a distorted view of the data, and they can appear fussy and take longer to interpret (see Figure 8.28).

Many visualisation tools are available within most standard desktop software and additional ones are available online. Some examples are:

- MS Office (Excel, Word, Visio);
- Mapping software (ArcGIS, QGIS, Google Maps);
- CAD/SketchUp.

No method is automatically excluded and innovative ideas may be welcomed. Students are advised to discuss their ideas with a supervisor. Some options are described below.

MAPS

A map is a 2D representation of an area of terrain, showing pertinent features and usually plotted to a scale. A map can readily communicate spatial information to the reader.

Maps may be used in hard copy format, or accurate Ordnance Survey maps for the UK may be downloaded by students for use in their studies through the Digimap site. Some maps and data sets to generate maps may be available online. These may be edited or annotated to allow key information to be highlighted. One example of this is Chloropleth maps, which are coloured or shaded according to relevant data.

Chloropleth maps can be created in GIS applications (e.g. ArcGIS) (Figure 8.28). Overlays can be done in Google Earth/SketchUp (see Figure 8.29).

3D MAPS AND PLOTS

Using 3D images can be a powerful means to communicate spatial data. These should be used with caution where the means of presentation can obscure important areas, a particular problem with any 2D representation of a 3D object.

These can be created using: mapping software (e.g. Google Earth) (Figure 8.29); CAD (e.g. AutoCAD); or SketchUp (Figure 8.30).

GANTT CHART

A Gantt chart is commonly used in project management to show which tasks are to be done at what time and how these relate to the overall project duration. The overall length of time allocated to each activity and where these overlap can also be seen (Figure 8.31). Using a Gantt chart to schedule dissertation tasks is highly recommended as it will readily communicate student intentions to the supervisor and identify where overruns of time are occurring. These can be created using MS Project, MS Visio (Figure 8.31) or MS Excel.

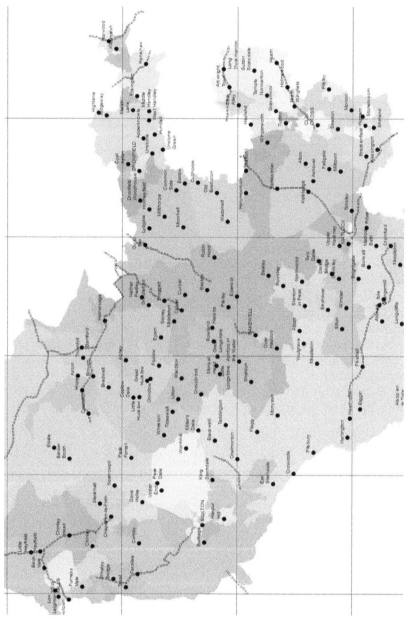

Figure 8.28 Chloropleth map

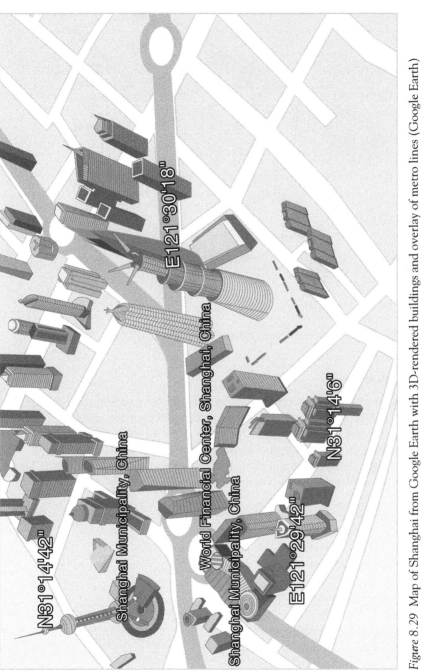

N31°14'42"

Shanghai Municipality, China

E121°30'18"

World Financial Center, Shanghai, China

Shanghai Municipality, China

E121°29'42"

N31°14'6"

Figure 8.29 Map of Shanghai from Google Earth with 3D-rendered buildings and overlay of metro lines (Google Earth)

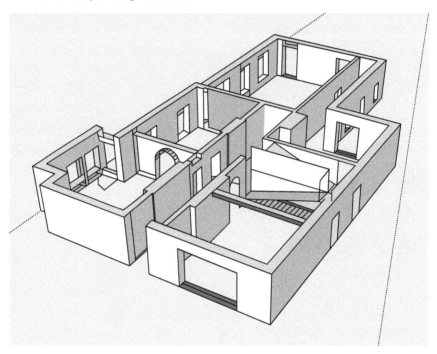

Figure 8.30 3D work in progress (created in SketchUp, 2015)

Statistical tests

Before any data are collected, a student should have an understanding of the likely nature of the data and the statistical tests that are suitable. The most important task is to make sure that there is an initial basic understanding of which tests are appropriate and what type (and quantity) of data is needed.

The use of statistics causes many students some concern, but there are lots of resources available to assist. These include:

- help functions within standard software (for example, MS Excel);
- free online resources, calculators and tutorials;
- textbooks on statistics;
- course tutors and university support.

The type of analysis will depend on:

- the number of variables:
  - univariate data – one variable;
  - bivariate – two variables;
  - multivariate – three or more variables;
- whether the data are parametric or non-parametric.

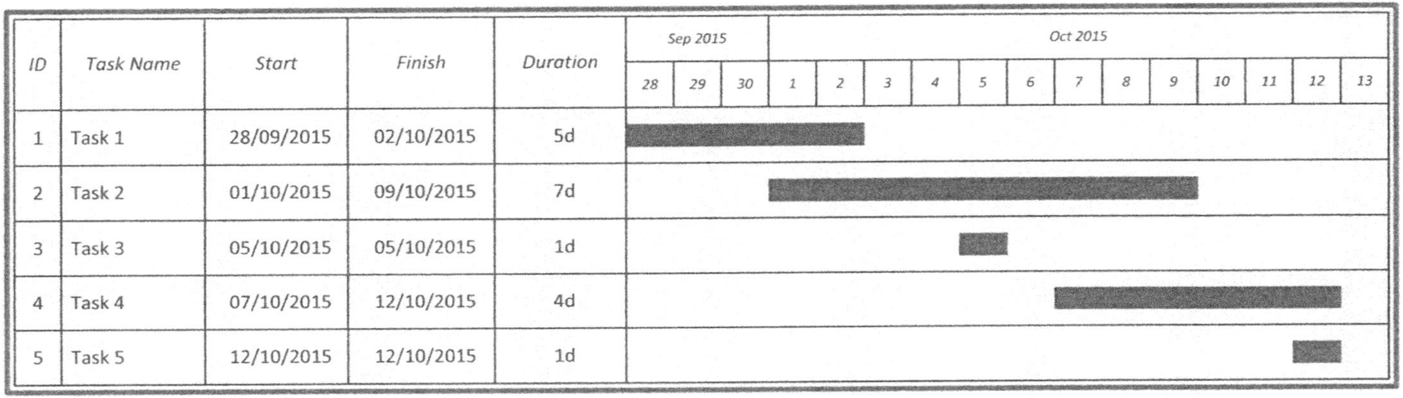

ID	Task Name	Start	Finish	Duration	28	29	30	1	2	3	4	5	6	7	8	9	10	11	12	13
					Sep 2015			Oct 2015												
1	Task 1	28/09/2015	02/10/2015	5d	▮	▮	▮	▮	▮											
2	Task 2	01/10/2015	09/10/2015	7d				▮	▮	▮	▮	▮	▮	▮	▮	▮				
3	Task 3	05/10/2015	05/10/2015	1d								▮								
4	Task 4	07/10/2015	12/10/2015	4d										▮	▮	▮	▮	▮	▮	
5	Task 5	12/10/2015	12/10/2015	1d															▮	

Figure 8.31 Gantt chart (created in MS Visio)

See 'Parametric and non-parametric data' in Chapter 6 for explanation of these data types. Some statistical tests require that data are parametric (underlying population is assumed to have a 'normal distribution' – see Figure 6.1) while other tests require non-parametric data (the underlying population does not approximate a normal distribution). Many parametric tests have non-parametric equivalents. Non-parametric tests are less powerful than their equivalent parametric tests. Where the data have low numbers of ranked scores, or a distribution with a few very high or very low numbers, then a non-parametric test should be considered. All tests involving ordinal (ranked) data are non-parametric.

Hypothesis testing

The student should understand the following terms in relation to the application of statistical tests in the testing of hypotheses:

- *Null hypothesis (H0)* – This states that there is no difference between sets of data, and that any apparent difference is just by chance.
- *Alternative hypothesis (H1)* – This states that there is a difference between sets of data other than could be caused by chance.
- *p-value* – This is a probability value which ranges from 0 to 1. *p*-values are used to establish if the means from two sets of samples are sufficiently different to conclude that they were taken from two separate populations with different means. This is done because it could be possible due to sampling alone to obtain different means from the same underlying population. Usually, a value of less than 5 per cent is taken, so if the *p*-value is less than this ($p < 0.05$), the data is significantly different, and H0 is rejected and H1 accepted. A calculated *p*-value of 0.05 would indicate that if you re-did the same procedure from the same population, you would expect the result to be larger only 5 out of 100 times.
- *Critical value* – Some methods of statistical analysis give a value that needs to be compared to figures contained in published tables of critical values. The critical value is the point beyond which the H0 is rejected, and this is determined from the level of significance (known as alpha, α). Alpha (α) can be taken as 0.05 (5 per cent). To use the critical values table the number of degrees of freedom (*df*) is needed, which is taken as a value one less than the number of data points. If the calculated value calculated by a statistical test is greater than the critical value given in tables, then H0 is rejected and H1 accepted.
- *One-tailed vs. two-tailed tests* – A one-tailed test should be used when it is possible to predict which group will have the larger mean before any data are collected. This may have been based on previous work, on literature or on expectations of movement in one way. A two-tailed test is used when a prediction of which group will be larger cannot be made beforehand. A two-tailed test looks for differences in either direction. If there is any doubt, use a two-tailed test.

The procedure for hypothesis testing is summarised in Figure 8.32.

Once the relevant procedure has been carried out, it is vital that the conclusion is stated. This is not just a case of writing the numbers down, but of stating the outcome in terms of the rejection or acceptance of the null hypothesis and what this leads the researcher to conclude about the data as a result. Conventions vary between disciplines and a student is advised to look at previous work and consult tutors to establish if there is a preferred convention at their home institution.

In stating the outcome of their work, students might use statements such as the following:

- HO is rejected because [insert calculated value] ≥ [value from table] at $\alpha = 0.05$, and it is concluded that there is a significant difference/change.
- HO cannot be rejected because [insert calculated value] ≤ [value from table] at $\alpha = 0.05$, and it is concluded that there is no significant difference/change.
- Since the p-value [insert calculated value] is less than the significance level (0.05), the null hypothesis is rejected, and it is concluded that there is a significant difference/change.
- The calculated value of [state test] is [insert calculated value]. This is greater/less than the [insert the critical value] [insert significance], therefore HO is accepted/rejected and it is concluded that ....
- The [state test] calculated value = [state value] and the p-value = [state value], leading to the conclusion that ....
- The [state test] resulted in a p-value of [state value] which is deemed statistically significant/non-significant, and therefore it is concluded that ....

A statistically 'significant' result is one where it would be unusual to find that the populations were identical given the observed differences in the data tested. This is not the same has having a result which is 'of importance', and a result where HO is accepted may be as or more pivotal to the research. To reduce the incidence of bias, all tests undertaken should be included when writing up the dissertation, not just those which are found to be 'significantly different'.

Brief overview of selected statistical methods

Choosing the correct statistical test is not always obvious. It is important to select appropriate tests before the data are collected at the point when methods and data collection are being considered.

When selecting statistical tests, a key question is: What is the intended purpose of the analysis? Is it to:

- describe the data (Table 8.5);
- investigating relationships between data (Figure 8.33);
- investigating differences between data sets (Figure 8.34)?

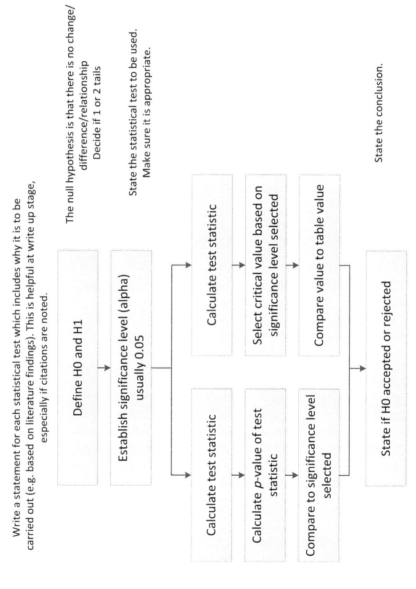

Write a statement for each statistical test which includes why it is to be carried out (e.g. based on literature findings). This is helpful at write up stage, especially if citations are noted.

The null hypothesis is that there is no change/difference/relationship
Decide if 1 or 2 tails

State the statistical test to be used. Make sure it is appropriate.

State the conclusion.

Define H0 and H1

Establish significance level (alpha) usually 0.05

Calculate test statistic

Select critical value based on significance level selected

Compare value to table value

Calculate test statistic

Calculate *p*-value of test statistic

Compare to significance level selected

State if H0 accepted or rejected

Figure 8.32 Flow chart for hypothesis testing

Table 8.5 Statistical methods of describing data sets

Aim of analysis	Type of data	Appropriate techniques
Compare two groups	Non-parametric: categorical	Chi square test compare the distributions
	Non-parametric: ordinal	Chi square test compare the distributions
	Parametric: Ratio/interval	t-tests, independent samples compare the means of two sets of data
Compare more than two groups	Non-parametric: Categorical/ordinal	Chi square test compare the distributions
	Parametric: Ratio/interval	ANOVA analysis of variance compare the means of more than two groups of data
Compare two variables over the same subjects	Non-parametric: Categorical/ordinal	Chi square test compare the distributions
	Parametric: Ratio/interval	t-tests, dependent samples compare the means of two sets of data

Correlation

Correlation quantifies how well the variables x and y vary together. The correlation coefficient (r) ranges from -1 to 1, where 0 means there is no correlation, $+1$ is a perfect positive correlation and -1 is a perfect negative correlation.

If r is close to 1 or -1, this could be due to one of four outcomes:

- x determines the value of y.
- y determines the value of x.
- Another variable influences both x and y.
- The correlation has happened by chance and there is no real correlation between x and y.

It should be noted that a correlation between variables does not always indicate that there is an underlying causation effect.

Pearson's r

Pearson's r (Pearson product-moment correlation coefficient) identifies correlations between scores for parametric data.

Pearson's r should be used where there are no outliers at either end of the distribution (if there are, use Spearman's rho). Data should be independent.

Results can be obtained using manual calculations, statistical software or Excel. The formula for calculating Pearson's r in Excel is:

=PEARSON(array1,array2)

For the data shown in Figure 8.35, Pearson's r is 0.97 showing a strong but not perfect correlation.

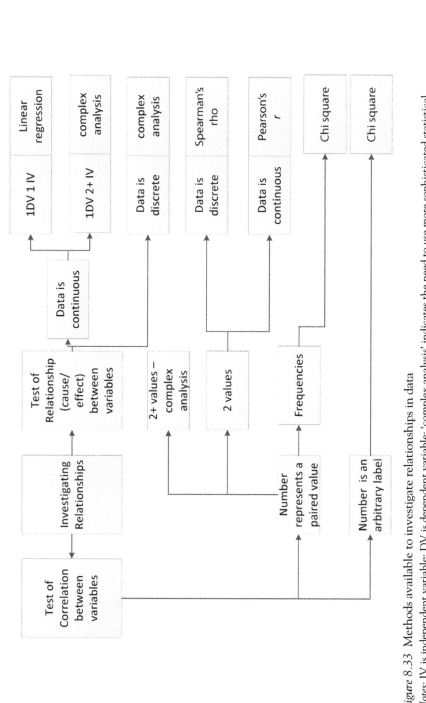

Figure 8.33 Methods available to investigate relationships in data

Notes: IV is independent variable; DV is dependent variable; 'complex analysis' indicates the need to use more sophisticated statistical test methods.

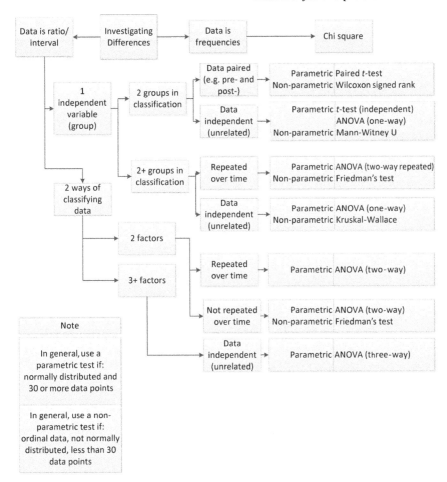

Figure 8.34 Methods used to investigate differences in data

Spearman's rho

Spearman's rho (Spearman's rank correlation coefficient) is the non-parametric test equivalent of the Pearson's r correlation. The relationship between x and y does not have to be linear. The test can be used with ordinal or ratio/interval data.

Spearman's rho may be used where there are outliers at either end of the distribution (if the data are normal, the Pearson correlation may be used). Comparing Spearman and Pearson correlations for the sample can give an indication whether outliers are influencing the results.

Results are obtained by: manual calculations, statistical software, online calculators or Excel (see Zaiontz, 2015).

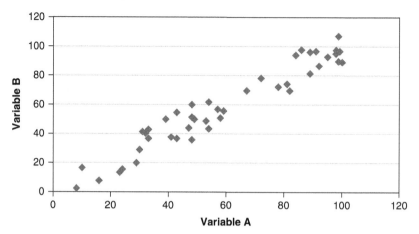

Figure 8.35 Scattergraph suggesting a correlation between variable A and B

Linear regression

Linear regression finds the line which best predicts *y* from *x* (or *x* from *y*).

The formula for a line can be inserted on a trend line in Excel (Figure 8.36).

It should be noted that a strong linear relationship between values does not always indicate that there is an underlying causation effect.

Chi square (X^2) test

The chi square test (X^2) is used with categorical or ordinal data to evaluate if observed differences in the distribution of categories within the data occurred by chance. It can be used to compare:

- two groups of categorical data;
- two groups of ordinal data;
- more than two groups of categorical or ordinal data.

The null hypothesis is that any apparent differences in distribution occurred by chance only.

The two comparisons are 'goodness of fit' and the 'test of independence'. Goodness of fit tests whether the distribution within categories matches a theoretical distribution. The test of independence considers whether the distribution within categories is affected by independent variables.

The more categories present, the more data are needed. As a rough guide, the minimum amount of data required before this test can be considered is (no. of rows × no. of columns × 5). For example, based on Table 8.6a it would be necessary to target 20 points of data before the test can be considered while, for Table 8.6b, at least 80 points of data would be required.

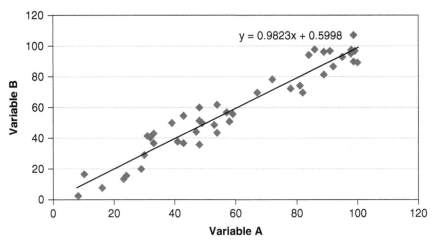

Figure 8.36 Trend line and equation on a scatter graph

Table 8.6a Estimating numbers of questionnaires for the chi square test, two rows and two columns

	C1	C2
R1		
R2		

Table 8.6b Estimating numbers of questionnaires for chi square analysis, four rows and four columns

	C1	C2	C3	C4
R1				
R2				
R3				
R4				

The 'expected' cell count must meet certain criteria – a general rule is: 5 or more in all cells of a two by two table; 5 or more in 80 per cent of cells in larger tables, but no cells with zero expected count. If these conditions do not apply, Yates' correction can be used (see Appendix).

Data should be independent (not matched pairs) and sampled randomly.

Check whether the output you are calculating is a value which needs to be referenced against a published table of critical chi square values or a p-value.

The limitations of this test are as follows:

- It does not give an indication of the strength or significance of the relationship.
- It is sensitive to large sample size.
- It is sensitive to small expected frequencies (expected values should be greater than five in each cell).

Results can be obtained from manual calculations, statistical software, online chi square calculators or Excel. A worked example is given in the Appendix.

t-tests

t-tests may be: one sample; independent; paired or dependent.

Data should be: interval or ratio; normally distributed; and randomly sampled.

This test compares means. The null hypothesis is that there is no difference between means. The alternative hypothesis is that there is a difference between means.

Whether or not the test will be one-tailed or two-tailed should be determined before starting.

A limitation of the *t*-test is that it examines means and not individual scores, so conclusions are general to the population and not to any individual point within the data set.

t-test (one sample)

A single sample *t*-test is used to compare the mean of a single set of data and a known or hypothetical population mean.

H0 is that there is no difference between the sample mean and the known (or hypothesised) mean (see the equation in Figure 8.37).

where:
t is the calculated t value.
μ is the expected value you are comparing against;
$\bar{x}$ is the average calculated from collected data;
s is the standard deviation of collected data;
n is the number of pieces of data

The following are also needed:
df – degrees of freedom $(n-1)$;
a – the level of significance (0.05).

Excel can be used to calculate the critical value:

=T.INV.2T(0.05,enter number of degrees of freedom)

This returns a two-tailed critical value of t.

=T.INV(0.05,enter number of degrees of freedom) –

$$t = \frac{(\bar{x} - \mu)}{\frac{s}{\sqrt{n}}}$$

Figure 8.37 t-test (one sample) equation

This returns a left-tailed critical value of *t*

=T.DIST.RT(0.05,enter number of degrees of freedom)

This returns a right-tailed critical value of *t*

If the calculated value of *t* is greater than the critical value, H0 is rejected and H1, that there is a significant difference, is accepted.

t-test (independent)

Use this to determine whether the means of two independent samples of parametric data have a different mean. The groups of data in this test should not be related in any way, they should have the same variance (see ANOVA), the data should be normally distributed and there should ideally be more than 30 scores.

Results are obtained from statistical software, online calculators and Excel. For Excel, the formula is:

T.TEST(array1,array2,tails,type)
Array 1 is the first group of data collected
Array 2 is the second group of data collected
For tails, enter 1 for a one-tailed test or 2 for a two-tailed test
For type, enter 1 for paired, 2 for two-sample equal variance, 3 for two-sample unequal variance

To determine if variances are equal or not, an *F*-test can be carried out with the formula:

=F.TEST(array1,array2)
Array 1 is the first group of data collected
Array 2 is the second group of data collected

If the *p*-value returned is less than 0.05 then the variances are different.

Paired t-test

Samples within a paired *t*-test are related. The paired *t*-test can be used to test for either a 'change' in these groups between two time points (repeated measures) or a 'difference' in the means of two related groups as the result of an intervention (matched-group). It cannot do both at the same time.

Since the data are considered as a pair, the degrees of freedom is the number of pairs.

Results may be obtained from statistical software or Excel. The following shows the process when using Excel:

1. Use the Analysis ToolPack.

2. Click the Microsoft Office button.
3. Select Add-ins to open a box.
4. Select Analysis ToolPack and click OK.
5. Select Data Analysis.
6. *T*-Test: Paired Two Sample for Means.
7. Select variable 1 range.
8. Select variable 2 range.
9. Alpha 0.05.
10. Select a cell to start output range (be careful not to overwrite any cells as this cannot be undone).

If the *p*-value given is less than the critical *p*-value then H0 is rejected and H1, that there is a significant difference in the means, is accepted.

ANOVA

ANOVA (ANalysis Of VAriance) can test hypotheses that the *t*-test cannot.

The groups of data should have the same variance (check using an *F*-test as described above for *t*-tests) and the group sizes should be approximately equal.

If a significant difference is found when the test is run, it does not identify which group varies or how, and post-tests must be run. Specialist statistical software may be needed for more complex ANOVA analyses.

A *one-way independent ANOVA* is used to determine whether the means of a single variable from three or more independent samples of parametric data are different; for example, a control group and two test groups.

A *one-way repeated measures ANOVA* is used to determine whether the means of a single variable from three or more repeated measures of parametric data are different; for example, a group measured at several time intervals or where each participant is placed in a number of different conditions and measured.

In both the one-way independent ANOVA and the one-way repeated measures ANOVA, if the means are found to be different, a post-test *t*-test can be carried out to find out which group(s) vary.

Results can be obtained from statistical software or using Excel. A worked example is given in the Appendix for a one-way independent ANOVA.

Kruskal-Wallace

The Kruskal-Wallis test is the non-parametric alternative to the one-way ANOVA test used to compare three or more samples. The null hypothesis being tested is that all populations have identical means. Data should be at least of ordinal level and the samples should be independent.

Results can be obtained using statistical software, online calculators or Excel (see workings in Zaiontz, 2015).

Wilcoxon signed rank

The Wilcoxon signed rank test is a non-parametric test that tests the differences in rank between paired measurements. Rejection of H0 leads to the conclusion that the populations have different medians.

Results can be obtained using manual calculations, statistical software or Excel (see Zaiontz, 2015).

Mann-Whitney

The Mann-Whitney test identifies whether the distributions of the two populations of non-parametric data are identical.

Results can be obtained using statistical software, online calculators or Excel (see Zaiontz, 2015).

Friedman's test

The Friedman test is a non-parametric test that compares three or more paired groups.

Results can be obtained using statistical software, online calculators or Excel (see Zaiontz, 2015).

Binomial test

The binomial test can be used to compare the statistical significance of an apparent deviation from the expected distribution when there are two categories. A large sample size and normal distribution is assumed. Population variance may be known or estimated from the sample. A sample size of 30+ is suggested, larger as p moves further from the range around 0.5.

The z test statistic is used

z 0.025 = 1.96 (critical value for two-tailed test)
z 0.05 = 1.64 (critical value for one-tailed test)

Results are obtained from: statistical software, online calculators or Excel (see Appendix).

Discussion and conclusion

Having discussed the analysis of data, this chapter goes on to consider how the results of analysis are used within the dissertation, focusing on the discussion and conclusion sections – the most important parts of the dissertation. The quality of the whole study will be demonstrated in these sections. The discussion section is used to convince the reader to accept the findings of the body of work. Every discovery, implication and consequence must be presented in a detailed, logical and readable manner. The reader must be left in no doubt as to the scope and nature of the entirety of the work and how this has been used. Conclusions demonstrate whether the aims of the work have been reached, the proof of which is in the main body of the dissertation text. The conclusion states, in the student's own words, the final results and the answer to the research question that was set at the beginning.

Writing the discussion

Before commencing the discussion section, a review of the process so far will be needed. All research projects are different, and therefore any approach must be tailored to meet the particular piece of work, but a general guide is provided in Table 8.7.

Many students see their research as a simple linear flow of events. In actual fact, for the majority of students, the decisions made as they go along will cause a backwards review of plans and moving forward in a slightly new direction. Each stage of the process can overlap, and undertaking the practical procedures of the whole process can be quite detailed and involved. It is important to address these shifts as the work is finalised because your aims and objectives may have shifted informally as your work developed. It is very likely that this will have been discussed with a supervisor but not formalised in the work. Writing a discussion and conclusion which reflects the work delivered without considering that this may not answer the initial aims and objectives as written several months previously is not advised.

A dissertation is a systematic investigation into a subject. In order to purposefully carry out the methodological aspects of the work, the student must have, through background reading and preliminary literature review, identified the key issues or problems within the subject of study. A student should demonstrate a good knowledge of the subject matter and methods that will enhance their work.

A diagram showing a framework for linking the early chapters with the discussion and conclusion sections is given in Figure 8.38.

Writing the conclusion

The conclusion is the final output of the research and should be written in such a way that the full findings of the project are summarised and presented to the reader. Conclusions link the aims and objectives with the data collected and the analysis in order to answer the original research question. Research may create a small number of main conclusions – perhaps up to a dozen. The transition from results to conclusions and recommendations through inferences requires insight – the conclusion should express and explain those insights such that the conclusions and recommendations will be informative. Formulation of conclusions may be guided by the following questions:

Table 8.7 Review of the research work undertaken and the choices made

Item	Focus	Link to relevant chapters
Summary of the problem	What was the purpose of the study? Which research methodology was chosen?	Chapter 2 Chapter 3
The literature review	What was already known about the subject? What information, or data, was required?	Chapter 5
Considering methods of investigation	What was the research basis: exploratory; descriptive; causal? What type of questions were asked? Was the principle source of the data primary or secondary?	Chapter 5 Chapter 6
Choices made in sampling (if carried out)	Who or what was the data source? What was the population? Was a sample necessary? Was the sample selected to be representative of the population? What level of accuracy was needed? How large was the sample? How was the sample gathered?	Chapter 6
Undertaking the investigation – collecting information or data	Was the information or data readily available? How was this measured? How long did data collection take? What procedures were used? Were the answers obtained?	Chapter 6 Chapter 7
Analysing the data	How was the data sorted and analysed? What level of analysis was done to satisfy the investigation? What quality control was there?	Chapter 7 Chapter 8
Assess reliability of work	Is it expected that the same results will be repeated if the research is repeated? Are the measurements used accurate and consistent? Would the same results be achieved by another researcher using the same instruments? Is the research free from error or bias (on the part of the researcher and the participants)?	Chapter 6 Chapter 7 Chapter 8
Assess validity of work	Was the method used to collect data appropriate and robust? OR Do the results obtained compare with the results obtained by another method elsewhere? Can the instrument used be correlated with pre-existing studies?	Chapter 6 Chapter 7 Chapter 8
Assess generalisability	Are the findings applicable in other research settings? Could the findings be applied to other populations?	Chapter 6 Chapter 7 Chapter 8

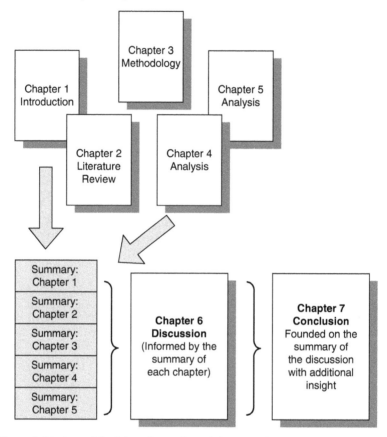

Figure 8.38 A model of the relationship of chapters, discussion and conclusion

- Have the data been interpreted correctly?
- Is it clear that the questions have been answered?
- Can the conclusions be linked back to the investigation?
- Are the conclusions supported by the data?

In summary, conclusions should:

- formalise the outcomes of the research, highlighting key findings;
- provide answers;
- be precise, comprehensive and have appropriate detail;
- have a clear focus;
- be truthful.

Recommendations and limitations

A short section following the conclusion may address limitations of the work done and recommendations for future research projects.

Limitations

Limitations explain why the scope of the study, the results or any other aspect were constrained. All undergraduate research (and arguably all research) has limitations, and recognition of these is important. Certain limitations are likely to have emerged during the course of the research; for example, not everyone who agreed to respond to a survey returned a completed questionnaire. In addition, circumstances change, favourably as well as unfavourably. Limitations can be set by the researcher or imposed on the research by changing circumstances. Research is dynamic and occurs in a dynamic environment, hence change is unavoidable. It is important to note both the type of limitation and the consequence.

Recommendations for further work

Where recommendations for further research are made, this should be based on an evaluation of the work undertaken. Consideration should be given to the methods appropriate for the work and a rationale for their selection. Mere recommendation of replicating the study with a larger sample that might reinforce the results is not a useful recommendation. Recommendations should ideally look to suggesting topics for future researchers based on the areas which the researcher felt had emerged during their work.

Supervisor guidance regarding data analysis and presentation

Guidance from Supervisor A

Careful consideration must be given to how you are going to analyse and present your data. When considering which techniques to employ for the analysis of collected data, clear reference must be given to your research objectives and the intended meaning of your research results.

If you are undertaking experimental laboratory research that seeks to collect quantitative data then your analysis will also be quantitative in nature. It is more than likely that various graphs and illustrations will support the presentation and explanation of your data.

If your research has taken place in a non-laboratory environment then you may have collected purely descriptive qualitative data, or possibly a combination of descriptive qualitative data together with numeric quantitative data.

When undertaking experimental research, a high level of control is exerted by the researcher upon variables within the environment. Within the built environment, this approach may be used to research and develop materials and technical products. Here, analysis may focus upon testing a hypothesis and seeking to identify, document and explain causal relationships between variables. Quantitative, numeric methods of analysis are predominant when experimental research is undertaken. A requirement of experimental research is that data collection, analysis and results are repeatable and generalisable.

When undertaking non-experimental research – little or no control is exerted upon that which is researched – then data collection and analysis could incorporate either or both quantitative and qualitative data. If your data collection methods include questionnaires, observations and interviews, be they within a case study investigation, an ethnographic study or a phenomenological inquiry, the analysis of data must still align with the objectives of the research and the intended meaning of your research results.

Non-experimental research may seek to produce results that can commonly be described as being either 'rich descriptions' or 'grounded theory'. Rich descriptions are commonly associated with investigation of culture, specific activities and practices, or peoples' attitudes and opinions. It is imperative that rich descriptions are presented in an organised and structured manner. Grounded theory seeks to generate theory that is considered as being emergent from or grounded in observations. Such theory may be 'specific' to the observed context (not readily generalisable to other contexts).

There are a good number of software packages available to help with the organisation and analysis of data. The coding, categorisation, and statistical analysis of data can be greatly assisted by the use of such software packages. Make sure that you investigate which data analysis software packages are available via your university.

Guidance from Supervisor B

In my view, one of the most common issues encountered by novice researchers is a lack of understanding about how the collected data can be presented and analysed effectively.

The first part of this process is organising the data effectively. I use Excel a lot so I will set up a worksheet with my original data that is never edited. All edits happen in a separate workbook so that if any cell is overwritten, I still have the raw data as a backup. Each analysis I do will be on a separate tab, and I tend to put the analysis at the top of the spreadsheet and the data below to save time in scrolling down sheets to find out the answers. Having lost data myself, I get very paranoid about backups and tend to save with a date in the file name to make sure that I don't overwrite any previous iteration of work.

All too often, students view software packages as a means of shortcutting the process. Where these are taught or supported, there is less of an issue, but there is a significant investment in time needed to teach yourself to use a specialist package. I would always advise that it is better to work with a system that you are comfortable and confident with. There is nothing fundamentally wrong with using little or no software. I would advise using some trial data to see how long things take you with different methods, and choose one which matches your personal requirements.

Choosing statistical tests always causes students a problem, and I understand that they may be frustrated when their supervisor does not sit down with them and instruct them exactly how to proceed. Students should ideally have a clear idea of the nature of the data and the types of analysis you might do before you collect the data. Basic descriptive stats can give you a quick 'way in' to your data. Take a strategic approach to which analyses you choose and only target ones which are linked to the literature or the aims of the test rather than adopting a scattergun approach and testing everything against everything. It is likely that you will find things of statistical significance, but they may be due only to chance. Another aspect to take care with is using numbers to represent categorical data. Be warned about looking for correlations in data where these arbitrary classes occur.

The tests most commonly used by my students are chi square, correlation (Pearson or Spearman), various types of *t*-test, ANOVA and the binomial test (for yes/no answers). SPSS isn't taught and only one or two students will use this, with the vast majority setting up spreadsheets to do their calculations.

I would like to see the standard of visualisations used in dissertations improve, as long as these are appropriate. I will say it here; the pie chart is one of the most commonly abused forms of data presentation I see. A pie chart that indicates that '100 per cent of respondents said yes' does nothing to improve the message or assist in the analysis. Where there are two responses divided on the basis of common fractions, such as a half or a third, then your supervisor will be able to picture this in their head without the need for a chart. A column chart is often a far more effective way of getting your message across. If you have long titles on your graph, use a bar chart so that the labels are clear. Many people will say that there is no one right way of presenting and analysing data and I completely agree. However, there are plenty of wrong ways of working.

Each visualisation needs a summary of the outcome in text; make sure that this does not simply describe the data presented in words. Instead, your text needs to communicate the outcome. The chart or visualisation needs to be referred to by number in the text. Don't only use the terms 'above' or 'below' to direct the reader to the relevant illustration as you may need to move visualisations around as you finally edit the document, which may result in the figures being somewhat separated from the text.

Guidance from Supervisor C

If a student has worked logically through the process to this point then their data presentation and analysis should be relatively simple. Students will have identified the analysis most appropriate for their data sets and will appreciate the vital difference between presenting their data and analysing it. Analysis should always be done bearing the aims and objectives of the research in mind to ensure relevance.

I personally despise 3D graphs as these are invariably unreadable. I instruct all my students to use only 2D graphs and to ensure that these are appropriately titled and referred to clearly in the text.

For questionnaire data, I urge students to take response rates into account. If only one questionnaire is sent and returned, this is a response of 100 per cent but by no means representative of appropriate research. I also work with my students to produce a Gantt chart and to finalise a date beyond which they will be undertaking analysis. At this date, any further responses are to be ignored. This strategy should, incidentally, also be adopted for secondary data as it is not possible to react at late stages to sudden changes in the real world; for example, a new law or change in government policy.

Student reflections on data analysis and presentation

'I found I wasn't able to measure and analyse my pilot questionnaire. Luckily I was able to change and re-do this and so didn't fall into the trap of not being able to analyse data. I would advise when choosing your topic try to think of something that might have had an effect on the given topic scenario or situation. This way you can measure the effect.'

'I had so much research material; at first, I didn't know how to use it the best way. I decided to file the evidence in date order and then by relevance. This worked for me.'

'Trying to find a way of presenting data effectively is really difficult sometimes. I used some free online graphics packages to create flow charts and floor plans and some online statistics packages.'

'I decided to use SPSS and found a really good online tutorial was available through the library. The only thing I couldn't get it to do was nice-looking graphs, so I ended up using Excel for those.'

'Deciding how to analyse interview transcripts was the thing I most struggled with. I read loads of textbooks which covered the theory but couldn't find a simple guide. In the end, I used markers to highlight blocks of text and then summarised that down into short blocks of text which were linked to each participant. I'm just glad I stuck with fully structured interviews or this would have been even more of a mare!'

'I used manual calculations in Excel for all my stats. It worked for me.'

'I found it really difficult to choose which graphs to use to show my data well. I really liked the look of the graphs I had produced. My supervisor hated them and told me to re-do all the 3D charts as they can be difficult for the reader to interpret. Reformatting every graph I wanted to use took me ages.'

Summary

This chapter has presented a range of approaches to data analysis and, in so doing, has outlined qualitative data analysis and its use of coding and categorisation in the undertaking of ethnographic, phenomenological and grounded theory studies; the use of quantitative techniques to illustrate and present qualitative data; and quantitative data analysis techniques for the analysis of numerical and statistical data.

Suggested further reading

ALLISON, B. & RACE, P. (2004) *The Student's Guide to Preparing Dissertations and Theses.* 2nd ed. London: Routledge Falmer. 0-415-33486-1
Part 2: Getting your act together, pp. 45–61.

BELL, J. (1993) *Doing Your Research Project.* 2nd ed. Buckingham: Open University Press.

BROWN, R. M. (2006) *Doing Your Dissertation in Business and Management: The reality of researching and writing.* London: Sage Publications Ltd. 1-4129-0351-3
Chapter 4: The problem with the research problem, pp. 37–50.

BRYMAN, A. & CRAMER, D. (1994) *Quantitative Data Analysis for Social Scientists.* Revised edition. London: Routledge.

BURNETT, J. (2009) *Doing Your Social Science Dissertation.* Sage Study Skills Series. London: Sage. ISBN 978-1-4129-3112-0

CLEGG, F. (1983) *Simple Statistics: A course book for the social sciences.* ISBN 978-0521288026

CRESWELL, J. W. (2012) *Educational Research: Planning, conducting, and evaluating quantitative and qualitative research.* 4th ed. Boston MA: Pearson. ISBN 978-0-13-261394-1

DAWSON, C. (2009) *Introduction to Research Methods. A practical guide for anyone undertaking a research project.* 4th ed. Glasgow: Bell & Bain Ltd. ISBN 978-1-84528-367-4

DENSCOMBE, M. (2010) *The Good Research Guide for Small-Scale Social Research Projects.* 4th ed. Maidenhead: Open University Press/McGraw Hill. ISBN 978-0-335-24138-5

DYTHAM, C. (2011) *Choosing and Using Statistics: A biologist's guide.* 3rd ed. Oxford: Blackwell Publishing.

FARRELL, P. (2011) *Writing a Built Environment Dissertation: Practical guidance and examples.* Chichester: Wiley-Blackwell. ISBN 978-1-4051-9851-6

FIELD, A. & HOLE, G. (2008) *How to Design and Report Experiments.* London: Sage, pp. 274–75.

FISHER, C. et al. (2010) *Researching and Writing a Dissertation: A guidebook for business students.* 3rd ed. Harlow, Essex: Pearson Education Limited. ISBN 0-273-71007-3

GILL, J., JOHNSON, P. & CLARK, M. (2010) *Research Methods for Managers.* 4th ed. Los Angeles; London: SAGE. ISBN 978-1-84787-094-0

GRAHAM, D. J. & MIDGELY, N. G. (2000) Tri-plot. Creative commons. Tri-plot original source. [Online]. Available at: www.lboro.ac.uk/microsites/research/phys-geog/tri-plot/index.html

GREETHAM, B. (2009) *How to Write Your Undergraduate Dissertation.* Houndmills, Basingstoke: Palgrave Macmillan. ISBN 978-0-230-21875-8

HERR, K. G. (2005) *The Action Research Dissertation: A guide for students and faculty.* Thousand Oaks, CA: Sage Publications Ltd. ISBN 0-7619-2991-6

HINTON, P. R. (1995) *Statistics Explained: A guide for social science students.* London: Routledge.

HOLT, G. D. (1998) *Guide to Successful Dissertation Study for Students of the Built Environment.* 2nd ed. Wolverhampton: Built Environment Research Unit, University of Wolverhampton. ISBN 1-902010-01-9

KVALE, S. and BRINKMANN, S. (2009) *Interviews: Learning the craft of qualitative research interviewing.* 2nd ed. Los Angeles: Sage. 978-0-7619-2542-2

MARDER, M. P. (2011) *Research Methods for Science.* Cambridge: Cambridge University Press. ISBN 978-0-521-14584-8

MELOY, J. M. (2002) *Writing the Qualitative Dissertation: Understanding by doing.* Mahway, NJ: Lawrence Erlbaum Associates.

NAOUM, S. G. (2007) *Dissertation Research and Writing for Construction Students.* 2nd ed. Oxford: Butterworth-Heinemann. ISBN 0-7506-8264-7

NATIONAL STATISTICS (2008) Results of the 2001 Census UK average population. [Online]. Available from: www.statistics.gov.uk/census2001/pop2001/united_kingdom. asp (accessed 1 August, 2015).

OAKSHOTT, L. (2009) *Essential Quantitative Methods: For business, management and finance.* Basingstoke: Palgrave Macmillan. ISBN 978-0230218185

OPPENHEIM, A. N. (1992) *Questionnaire Design, Interviewing and Attitude Measurement.* (New ed.). London: Continuum.

PIANTANIDA, M. and GARMAN, N. B. (1999) *The Qualitative Dissertation: A guide for students and faculty.* Thousand Oaks, CA: Corwin Press. ISBN 978-0803966895

ROBSON, C. (2011) *Real World Research.* 3rd ed. Hoboken, NJ; Chichester: Wiley. ISBN 978-1-405-18240-9

ROWNTREE, D. (2000) *Statistics Without Tears: An introduction for non-mathematicians.* London: Penguin. ISBN 978-0140136326

RUDESTAM, K. E. & NEWTON, R. R. (2007) *Surviving Your Dissertation: A comprehensive guide to content and process.* Los Angeles: Sage. ISBN 978-1-4129-1679-0

SMITH, K., TODD, M. & WALDMAN, J. (2009) *Doing Your Undergraduate Social Science Dissertation* [ELECTRONIC BOOK]. London: Routledge. 978-0415467490

SWETNAM, D. (2001) *Writing Your Dissertation: How to plan, prepare and present successful work.* 3rd ed. Oxford: How To Books Ltd. 1-85703-662-X

WALLIMAN, N. (2011) *Research Methods: The basics*. Abingdon: Routledge. ISBN 978-0-415-48994-2

WONNACOTT, T. H. & WONNACOTT, R. J. (1990) *Introductory Statistics*. 5th ed. New York: John Wiley & Sons.

YONG TECHNOLOGY INC (2014) GeoRose. Edmonton, Canada. [Online]. Available at: www.yongtechnology.com/download/georose

ZAIONTZ, C. (2015) *Real Statistics Using Excel*. [Online]. Available at: www.real-statistics.com (accessed November 2015).

9 Completing the journey

Writing up and assessment

Introduction

This chapter is concerned with writing up the dissertation and assessment. As such, it covers:

- the structure and contents of a final dissertation thesis;
- writing up and considerations for enhancing dissertation writing style;
- plagiarism and academic misconduct;
- the *viva voce*, the oral assessment of the dissertation – when it might be carried out, how to prepare for it and typical questions that might be asked;
- assessment of the dissertation proposal and the final dissertation thesis, including various examples of assessment criteria for the proposal and final thesis;
- guidance from three supervisors concerning writing up and assessment;
- some brief student reflections regarding writing up and assessment;
- suggested further reading.

Writing up

The dissertation enables students to demonstrate academic and research skills. These skills are shown in terms of delivering a research project and in the ability to present and discuss results in a suitable and meaningful manner.

Writing up a dissertation often presents a number of challenges. One challenge commonly encountered is writing the dissertation in a suitable style. It is often assumed by students that the frequent use of highly embellished language or technical terminology is the expected norm. This is not the case. Dissertation researchers should instead focus upon writing clear and precise English. Writing in such a manner can help to demonstrate clarity of thought and ensures that the finished work is readable.

Since the dissertation is often the largest single document produced by a student during their university life, there is an enhanced need for a systematic approach to organising and producing the thesis. Poorly expressed writing and unclear structure can certainly have an adverse effect on the reader's perception of the quality of the research work.

Layout of the dissertation thesis

The standard format for presenting a dissertation varies between universities. As such, all dissertation students are advised to verify the dissertation format required by their own university – this will be the format that must be utilised. Institutional conventions may require the use of certain fonts and font sizes. They may also specify margin sizes and general page and document layout. It is also common for specific referencing conventions to be specified. These requirements are not universal and do vary from university to university.

An indicative structure for a dissertation thesis is outlined in Table 9.1. This is sometimes referred to as the 'architecture of the dissertation'. Each new section outlined within Table 9.1 commences on a new page.

Table 9.1 Indicative order for the contents of an undergraduate dissertation

Section	Additional information
Title page	Full title of dissertation
	Author's name in full
	Title of degree
	Name of university
	Date (year and month) of submission
Abstract	A summary of the entirety of the work
	About 200 to 300 words long and should be on a single page
Contents	Sections and subsections with page numbers
List of figures	Full captions and page numbers
List of tables	A reference should be provided if taken from a published source
List of equations	
Acknowledgements	Family and friends, supervisor
	May also include a list of persons contacted and organisations who have provided information as long as consent has been obtained to reveal this information
Declaration	Wording as institutional requirements or, if there are none specified, use:
	I declare that the work contained within this dissertation is my own work and that no part has been plagiarised. Where work and theory or concepts have been taken or adapted from other authors, these have been properly cited or referenced. This dissertation stands at _____ words approximately.
	Signed:
	Print name:
	Date:
Nomenclature	Define acronyms or technical terms
Introduction	An overview of the work
Aims and objectives/ hypothesis	Identify the research focus
Main body	Broken down into related chapters presented in a logical order – in general terms, this covers the literature review, research methodology, and results
	Each chapter starts on a new page
Discussion	Bring together all strands of the research and examine what the findings mean

Table 9.1 (Continued)

Section	Additional information
Conclusion	Refer to the original aims and objectives/hypothesis Summary of key findings and their implications How findings fit with the existing knowledge presented in the literature review Limitations – reflect on the work produced and suggest where this could be improved Recommendations for further study
References	Sources used in the dissertation (may be titled Bibliography in some institutions) Alphabetically sorted by author surname
Bibliography	Sources related to but not used in the dissertation (may be titled References in some institutions) Alphabetically sorted by author surname
Appendices	Will normally include: • ethics pro-forma or assessment • examples of data collection instruments • summary sheets of (anonymised) data • log of work

Improving writing style

Presenting a dissertation requires effort to be invested into producing well-crafted, readable text. Ideally, the flow of words should carry a reader through the work. In order to achieve this, there should be minimal spelling mistakes and few grammatical errors. Frequent errors in spelling and grammar and the inappropriate use of dialect or badly constructed phrasing can distract the reader and negatively impact the perception of the work. As such, when writing up the dissertation, it is necessary to give good consideration to the style and structure of your writing.

There follows a concise overview of various aspects to consider when writing up.

Good writing style

- Use clear, concise English.
- Do not use an informal style.
- Do not write as you speak.
- Avoid unnecessary clichés, jargon, buzzwords or slang.
- Avoid double negatives.
- Unless advised otherwise, avoid writing in the first person. For example, use 'the interviews were carried out … ', not 'I did the interviews … '.
- Sentence construction should be precise.
- Do not be afraid of breaking down excessively long sentences into two or more sentences.

Spelling

- Use a spell checker (set to UK English).
- Check unfamiliar words using a dictionary.
- Check with a thesaurus to make sure that any unfamiliar words are correctly used.

Grammar

- Check for agreement.
- The term 'data' is plural so use 'the data are … '.
- Use appropriate punctuation and abbreviation (see Table 9.2).

Numbering

- The title page does not have a page number.
- Page numbers before the Introduction should be in roman numerals (i, ii, iii, etc.).
- Page numbers start at '1' from the Introduction.
- Numbering of chapters and section headings needs to be consistent.
- Use no more than four levels of numbering in section headings.

General good practice

- Follow the style guidelines of your home institution (e.g. for presentation, referencing and citation, etc.).
- Use the specified font and font size (e.g. Ariel or Times New Roman; 11 or 12 point).
- Use the specified line spacing (e.g. 1.5 line spacing).
- Never hand in a document without proofreading it.
- Keep a close track on the word count and compare to the word limits which were given.
- Use an outline to help organise sections.
- The document should make sense to a non-specialist.

Building paragraphs

When writing up, it is worth giving thought to paragraph structure and content. The paragraph can be considered a building block when presenting and discussing research work. Each paragraph should be a collection of linked or connected ideas. The opening sentence should direct readers to the focus of the paragraph. Try to avoid opening a paragraph with link words such as 'furthermore' or 'however'. When undertaking a literature review, try to avoid starting multiple paragraphs by citing particular works, as in '*White (2011) found that … *'. Focus instead on grouping similar themes, and construct a sentence which introduces the theme to be discussed in the paragraph. Citations will then form part of the main paragraph body.

Table 9.2 Punctuation revision

Item	Use
Capital letters	Use for people, places or titles
Comma	After an introductory phrase
,	To separate items in a list
	To separate clauses in a sentence
Semicolon	Used where a break is needed that is stronger than that provided
;	by a comma
	Where two complete sentences are linked; they can be joined
	with a semicolon (use sparingly)
Colon	Use before a list of bullet points
:	Use before a statement offering an explanation
Full stop	To end a sentence that is not a question
.	With initials and most abbreviations (there are exceptions to this)
	'e.g.' 'ed.' (editor) 'vol.' (Volume) 'no.' (number)
Apostrophe	Indicate possessions (the book's aims are …)
'	To indicate removal of letters (it is – it's)
Parentheses	Insert additional information into an already complete sentence
( )	(this would include citations)
Hyphen	Used in compound adjectives and some compound nouns
‐	
Dash	Indicates a range of values (e.g. 20–30)
–	
i.e.	*id est* or 'that is'
e.g.	*exempli gratia* or 'for example'
etc.	*et cetera* or 'and other things'
et al.	*et alia* 'and others' (people)

The main body of the paragraph needs to clearly set out the main focus with evidence to support the arguments. These may be substantiated with reference to tables, figures or other illustrations such as diagrams, but only where these add to the value of the argument. Consider the length of the paragraph carefully. One that has become too long will be unwieldy. However, breaking the paragraph clumsily into two or more sections will interrupt the reader's flow. A reasonable length for a paragraph is perhaps between 100 and 200 words. Each paragraph should be concluded by a single sentence, pulling the relative points together and making it clear what has been learned from the argument presented. This should link forward to the next paragraph where needed.

The document needs to be structured so that the reader is guided through the author's thought process in a systematic way. Appropriate linking words (Table 9.3) can be used to guide the flow of thoughts and ideas.

Hand in

Prior to hand in, students are advised to:

• Check and double check hand-in dates.

Table 9.3 Linking words

Purpose	Suggested linking words
Attributing a cause	as, for, since
Showing an effect	accordingly, as a result, for this reason, so, therefore
Intensifying meaning	as a matter of fact, in fact, indeed
Showing equality	equally, in the same way, likewise, similarly
Showing time relationship	after, before, once, until, when, while
Making a suggestion	if so, in that case, otherwise, that implies
Where a condition exists	as long as, even if, if, provided that, unless, whenever
Making a list	this includes, first, furthermore, in addition, lastly
Moving to a new subject	incidentally, regarding, with reference to, with respect to
Summarising a section	in brief, in conclusion, overall, to conclude, to summarise
Explaining something in a different way	notably, or rather, that is to say
Giving an illustration	for example, for instance, especially, in particular
Point clarification	in other words, rather, to clarify
Suggesting an alternative	alternatively, on the other hand, rather
Showing a contrast	although, conversely, despite, however, instead, in comparison, nevertheless, though, on the contrary, on the other hand
Where something unexpected was found	although, despite, even if, nevertheless, nonetheless, still, though, yet

- Check a draft to ensure that the content is in the correct order, and in the correct format, and that it is clearly referenced.
- Verify binding requirements and document format as well as the number of copies required.
- Identify any other requirements (e.g. CD copies, submission to anti-plagiarism software, emailed copies).

Plagiarism and poor conduct

In simple terms, plagiarism is taking the words or the ideas of others and using them without giving credit. At best, it demonstrates poor academic practice and, at worst, the theft of the work of others. Research integrity should form a backbone to practice, and this includes both honesty and openness throughout the process.

Referencing is important as it:

- provides a means of checking student assertions;
- demonstrates an understanding of the current state of knowledge;
- allows future researchers to more rapidly review the existing body of knowledge.

Researcher questions
Writing up

You should be aware of the requirements for the written document in terms of style and layout. You should have read the section 'Writing up'.

Yes. I have read this section and am ready to continue.

No. I have not read this section. I will review the requirements of my home institution and return to this chapter later.

Whenever a piece of existing literature is used, a student should ensure that a clear reference is made to the source ideas by giving an academic reference (Table 9.4).

The four Rs of research are:

Read it; then either *Repeat* it (quote) OR *Rewrite* it (paraphrase); and ALWAYS *Reference* it.

Avoiding plagiarism

Plagiarism is where the work of another, which may include thoughts or ideas, is being passed off as the student's own, whether deliberately or unintentionally, without appropriate acknowledgement.

Plagiarism can take a number of forms, including:

- copying the whole of or a substantial part of the work of another without acknowledgement of the source (includes copying from the work of another student, from existing creative works or from a book or website);
- inclusion of a small amount of another individual's work without it being referenced, including where a few words have been altered;
- duplication of a student's own work without acknowledgement where this has already been assessed for another part of a course or published elsewhere.

Plagiarism is a very easy trap in which to fall, but it can be avoided by maintaining a careful distinction between the ideas of others and the ideas of the student author. This is communicated to the reader by referencing.

Anti-plagiarism software

Anti-plagiarism software works by comparing student work to pre-existing work in a number of databases. The overall match is given by a score; however, this

Table 9.4 When to give a reference

Item	Needs a reference?	Notes
Text directly quoted from written source	Yes	Make sure that the quote is formatted properly according to the standards of the home institution. This may require speech marks, italicising and indentation of large blocks of text.
Images used from an existing source	Yes	A caption should be provided with the image. This may be titled as a 'Figure' with a sequential number followed by a descriptive title and the source reference.
Table of data from an existing source	Yes	A caption should be provided with the table. This may be titled as a 'Table' with a sequential number followed by a descriptive title and the source reference.
Paraphrased text	Yes	Paraphrased text is written by the student based on the ideas from one or more sources. Each of these sources should be clearly referenced so that the reader can find the source of the information in the original document to understand the foundation upon which arguments are based.
Case studies (published)	Yes	If a case study is published then it should be referenced if included in student work.
Case studies (undertaken)	(depends)	If existing data are used to assemble a new case study then this needs a reference unless the data are of a sensitive nature. Appropriate acknowledgement should usually be made.
Interviews undertaken by the student	(depends)	The presumption in an interview for a dissertation is that the participant's identity will be anonymous unless they have given their express permission to be named. It is recommended that interviewees be referred to by a letter or number and not named within the document; they should not be named in the reference list.
Table of data/image produced by the student	No	If a student wishes to show that they are the originator of content where there could be doubt on whether the material is primary or secondary in nature then they may add 'Source: Author' at the end of the caption. The exception to this is work that the student has already submitted for assessment or which forms part of an existing published work; in this case, the usual (Author, Date) citation is required with a corresponding reference.

is not the most important value when deciding whether plagiarism has occurred and the following are of relevance:

- the number of individual matches;
- the percentage match from each of these individual matches;
- how these matches have been used and referenced;
- where these matches are from (websites, wiki sites, blogs);
- the nature of the work.

Where work requires large amounts of text from existing sources to be quoted in the student work then the score will be high, but not of concern. Where there are numerous incidences of header and footer text which match pre-existing work, this may result in a high score. A rule of thumb is that an overall match of 30 per cent is viewed as a potential point of concern; and if the document has a match of 10 to 15 per cent (or above) to a single source then further investigations may be needed. Multiple small matches of less that 1 per cent are usually of lower concern.

Academic misconduct

Academic misconduct is most closely linked to failure to conduct work in an ethical manner. It may include the following:

- *misrepresentation* – for example, of data (presenting a flawed interpretation of data);
- *falsification of data* – for example, presenting experimental results which have been fabricated rather than recorded;
- *collusion* – this is where a student undertakes work with others where the work should be individual;
- *mismanagement* – failure of a student to act properly in the undertaking of research, including acting irresponsibly in terms of the data gathered and the health and safety of themselves and others as well as failure to adhere to ethical principles;
- *cheating* – including allowing another student to access their own work for the purposes of copying sections, taking unauthorised material into a *viva* examination (check requirements with the home institution), and plagiarism;
- *dishonest practice* – covers any form of dishonest practice, including but not limited to actions such as actual or attempted bribery.

The most obvious example of academic malpractice is for a student to buy a dissertation from another. The clear advice here is *Do Not Even Contemplate Doing This*.

The penalties for academic malpractice at final year of degree are severe and a student may face expulsion from their course without any academic credit. Plagiarism software, such as Turnitin, is not the only way that content is checked. For example, from the student's audit trail, the supervisor often has the ability to verify any of the process that has been carried out. There may be additional institutional verification of data collection or process. A student contemplating purchase of a dissertation from any source is advised that, even where there is an offer of a refund should the deception be uncovered, this will almost certainly not include compensation for the time and fees that were invested in the degree overall.

Researcher questions
Plagiarism and poor conduct

You should be aware of generic advice on plagiarism and poor conduct and how to avoid these, as well as the specific requirements of your home institution in terms of academic conduct.

Yes. I have read this section and am aware of the specific requirements of my home institution with reference to expected academic conduct. I am ready to continue.

No. I have not read the documentation. I understand that I need to have an understanding of the framework for good academic practice and that this will include understanding the requirements of my home institution.

It is advised that you read them before you continue to the next section.

The *viva voce*

Some universities may not use this method of assessment. If the dissertation at your home institution is not assessed by a *viva voce, you may want to skip this section.*

The *viva voce* (often shortened to *viva*) is an examination conducted orally rather than in writing. The purpose of the *viva* is to:

- assess your progress and understanding of the research process at different stages of your dissertation;
- support you in your development.

If done at the end of the dissertation process, it is used:

- to allow a student to defend their choices;
- to ensure that the work produced was understood;
- to question more deeply any aspects not covered;

- to verify facts presented;
- to demonstrate that all of the work is the student's own.

If done as an intermediate part of the dissertation process, it is used:

- to allow a student to present their initial views on direction;
- to verify student understanding of the process and rationale for ongoing work;
- to question choices in order to encourage reflection;
- to provide summative feedback on the process which may result in changes to the work;
- to demonstrate developing knowledge and understanding.

The format and time allocated for a *viva* will vary considerably between institutions. Timings can be fixed or open-ended. Some defences of work may be conducted in an interview setting, others in a presentation setting.

Guidance for viva voce

- Ideas should flow from the student without a great deal of prompting – but a student monologue should also be avoided.
- Listen to the questions carefully and ensure that the answers given address these.
- Answers should not be rushed.
- Use your enthusiasm for the topic and your imagination to give good, detailed answers.

Example 1: Indicative viva questions

When a *viva voce* is conducted at the end of your dissertation research, the following may be indicative of the types of questions that the student researcher might face:

Topic

- discussion on why the topic was selected;
- identification of the research outcomes (that were expected and that have been achieved);
- details of whether the research is novel or the aspects which may have an element of originality;
- explanation of the nature and type of the limitations imposed on the study;
- identification and discussion of key texts.

Methods of research

- research methodology;
- aims/objectives/hypothesis;

- recognition of the choices of research method that were available;
- techniques used;
- the forms of data which result from methods used (qualitative and quantitative);
- the timescales and limitations;
- the use of contingency plans;
- alternative approaches and the rationale for rejection of any methods;
- advantages and disadvantages of methods chosen.

Methods of analysis

- recognition of the types of analysis that were available;
- the type(s) of data analysis used and their appropriateness for the data collected;
- strongest/weakest parts of the analysis.

Area of research

- topics aligned to or overlapping the area of study;
- the relevance of the work.

Example 2: Indicative viva questions

When a *viva voce* is conducted part way through your dissertation research as a means to refine your research proposal and develop your work, the following may be indicative of the types of questions that the student researcher might face:

- Outline what your chosen research topic is, and why have you chosen it?
- Can you please talk us through your research aims and objectives?
- Are these realistic and achievable or could they be considered to be a little ambitious?
- What key sources of information have you drawn upon to help develop your knowledge and understanding of the topic of this research investigation?
- How do you intend to gather data to achieve your research aims and objectives/test your hypothesis?
- Can you please explain to the examiners what data collection and analysis methods you have selected and why?
- Do you envisage any difficulties in carrying out your research?
- Are there any external constraints that could prevent you from achieving your objectives?
- Are there any significant ethical issues associated with your research?
- Is there anything that you would like to ask the examiners?

Preparation for viva voce

- Ensure aims, objectives and methodology are clearly understood and can be articulated and explained.
- Make preparations to explain and justify the methodological model.
- Identify 'weak spots' in the work (items which haven't been or won't be done).
- Start a file of anticipated *viva* questions.
- Be familiar with the references used.

Some institutions allow written notes to be brought in. If this is allowed, then:

- Keep these to an absolute minimum.
- Create a brief summary of work with index tags so items can be found quickly.
- Minimise the time spent reading from the written material. Maintain eye contact with audience.

There may be questions that are difficult to answer, in which case:

- Ask for clarification on any question where the meaning is unclear. One way of doing this is to rephrase the question as 'Do you mean … [paraphrase the question]?'
- If there appears to be a misunderstanding on the examiners' part about the work, try to isolate this and give an explanation.
- If a question arises that cannot be answered – be honest.
- If mention is made of literature that has not been part of the student review then ask the examiner if they would provide you with the reference after the *viva*.
- If any aspect is forgotten, ask if that can be revisited later in the *viva*.

Do not panic. The purpose of a *viva* is to put a student in an examination situation, which is often seen as stressful. Be calm, talk clearly and remember that you, the student, should be the one most familiar with the work in contrast to the situation in any other type of examination.

Assessment criteria

Dissertation assessment criteria vary from university to university. Dissertation researchers are therefore advised to locate and review the assessment criteria used by their own university and degree course.

> ## Researcher questions
> ### The *viva voce*
>
> You should now have an understanding of the requirements of the *viva voce* (and you should have checked whether your dissertation will be examined in this way).
>
> **Yes.** The assessment of my dissertation requires a *viva voce* to be carried out. I have read the relevant section in conjunction with any guidance provided to me by my university. I understand what is expected from a *viva voce*.
>
> **No.** The assessment of my dissertation does not require me to be orally examined with a *viva voce*. I have doubled checked this with my university.

Examples of assessment criteria

The assessment criteria provide the mechanism for the grading of the dissertation. It is essential that dissertation researchers familiarise themselves with the assessment criteria that are to be applied to their dissertation. A number of indicative assessment criteria marking schemes are outlined and presented within this section. These demonstrate various approaches to assessment with different levels of detail. Included here are three examples relating to the research proposal, five examples for the final dissertation, and one for the *viva voce*.

> ## Researcher questions
> ### Examples of assessment criteria
>
> Before reading this section, you should have a clear understanding of the tasks which will need to be completed according to your home institution.
>
> **Yes.** I have a clear understanding of the assessment criteria specified by my university. I also have an understanding of what I am required to submit for assessment and when.
>
> **No.** I am not aware of the assessment criteria specified by my university. I do not yet have a clear understanding of what I will be required to submit and when. I realise that I will need to have this information in order to continue my journey and succeed in the dissertation challenge. As such, I shall now undertake to determine the assessment and submission requirements of my university. After completing this task, I shall return to read this chapter.

Research proposal: example marking scheme 1 (list of criteria)

Aims, objectives and hypothesis

- Aims clearly stated
- Objectives clearly stated
- Appropriate linkages shown (e.g. methodological model/hypothesis)

Research methodology

- Theory clearly stated
- Reasons for selection of methods given
- Aims and objectives match methodology
- Methodology appropriate for proposed research
- Evidence of deeper consideration of research question
- Ethical review including Risk Assessment

Depth of knowledge of subject area and content

- Good range of references clearly cited
- References from a variety of sources (Internet/journal/book)
- References appropriate to subject
- Consideration of validity of each of sources/critical analysis
- Content reflects understanding of research question

Marker guidelines: Consider range of references and quality. A 2:1 or higher proposal must have accurate citation in the text.

Conclusions

- A well-rounded conclusion drawn from issues presented
- Suggestion for further study/identifies areas of uncertainty

Presentation

- Clarity of expression
- Spelling and use of English
- Standard of presentation
- Appropriate referencing system (reference list and citation)

Marker guidelines: Appropriate presentation. Thoughts and ideas clearly expressed. Grammar and spelling accurate. A 2:1 or higher proposal must have an accurate and appropriately formatted reference list.

Table 9.5 Example marking grid for research proposal

Criteria	1st class (70%+)	2.1 (60–69%)	2.2 (50–59%)	3rd class (40–49%)	Refer (0–39%)
Presentation and clarity of expression	Presentation shows an imaginative approach to the topic. Thoughts and ideas are clearly expressed. Grammar and spelling accurate. Fluent academic writing style.	Presentation carefully organised. Thoughts and ideas clearly expressed. Grammar and spelling accurate and language fluent.	Presentation satisfactory, showing some organisation and coherence. Language mainly fluent. Grammar and spelling mainly accurate.	Presentation shows an attempt to organise in a logical manner. Meaning apparent but language not always fluent. Grammar and spelling contain errors.	Presentation is disorganised/incoherent. Purpose and meaning of dissertation and/or language unclear. Grammar and spelling contain errors.
Aims/ objectives	Clear aims and objectives that are well linked and outcomes clearly defined.	Aims and objectives linked and outcomes stated.	Aims and objectives produced and linked but outcomes not clear.	Aims and objectives produced but outcomes not clear.	Aims and objectives not linked and no defined outcomes.
Research methodology	Methodology allows the completion of the aims and objectives. Also triangulation of results is employed.	Methodology will enable the completion of aims and objectives.	Methodology selected will allow most of the aims and objectives to be completed.	Methodology selected will only allow partial completion of aims and objectives.	Methodology employed will not enable the completion of the stated aims and objectives.

Table 9.5 (Continued)

Criteria	1st class (70%+)	2.1 (60–69%)	2.2 (50–59%)	3rd class (40–49%)	Refer (0–39%)
Use of literature and referencing	Has developed and justified own ideas based on a wide range of sources, which has been thoroughly analysed, applied and discussed. Referencing is consistently accurate using the Harvard or numerical system.	Able to critically appraise the literature and theory gained from a variety of sources; developing own ideas in the process. Referencing is mainly accurate using the Harvard or numerical system.	Clear evidence and application of literature/theory relevant to the subject. Indicative texts identified. An attempt at referencing has been made using the Harvard or numerical system, but this lacks detail.	Literature is presented uncritically and indicates limitation of understanding. Literature has been used in a purely descriptive way. Some referencing has been incorporated but not using the Harvard or numerical system.	Literature either not consulted or irrelevant to dissertation. Referencing is absent and this may suggest plagiarism.
Conclusions	Analytical and clear summary that is well grounded in theory/literature and the student's own investigations, showing development of new concepts.	Good development shown in summary of arguments based in theory/literature and the student's own investigations.	Evidence of findings and summary of arguments grounded in theory/literature and the student's own investigations.	Limited evidence of findings and summary of arguments supported by literature/theory and the student's own investigations.	Unsubstantiated/invalid summary of arguments based on anecdotes and generalisations only. No real investigations undertaken by student.

Research proposal: example marking scheme 2 (grid). The example marking grid is given in Table 9.5

Research proposal: example marking scheme 3 (summary)

Title
Working title stated in a maximum of eight words.

Specification of the problem
Identify the problem underlying the research. Identify the single solution which will be the aim of the work produced.

Literature review
Provide a critical review of the literature demonstrating clear knowledge of the area. At least 12 sources should be used.

Methodology
State the research question or hypothesis. Justify the theoretical framework and methods to be used.

Ethics approval
Give all forms and relevant information.

Final dissertation: example marking scheme 1 (list of criteria)

Aims, objectives and hypothesis

- Stated clearly
- Representative of content of document

Depth of knowledge (use of literature and referencing)

- Range of references appropriate to subject
- References from a variety of sources (Internet/journal/book)
- References analysed/demonstrates understanding of topic area
- Consideration of validity of each of the sources/critical analysis
- Referencing in main text accurate to Harvard or numerical system

Marker guidelines: Appropriate presentation. Thoughts and ideas clearly expressed. Grammar and spelling accurate. A 2:1 or higher dissertation must have an accurate and appropriately formatted reference list.

Data collection and analysis

- range of data collection
- presentation of data using appropriate methodologies
- analysis of data using appropriate methodologies

- findings expressed in an appropriate and rigorous way
- evidence of deeper consideration of research question and/or advanced interrogation of data
- research ethics statement presented in the appendices.

Marker guidelines: Range of data collection. Extent and appropriateness of analysis of data. Identification of patterns and themes within the data collected. Critical analysis of data for reader. Where numerical analysis provided, consider range of appropriate graphs and charts used to illustrate data. Statistics should be used to analyse numerical data. Where interviews and qualitative methods used, thorough and critical analysis.

Discussion and conclusions

- Discussions grounded in theory/literature
- Triangulation of data sources evident
- Summary uses student findings in appropriate manner
- Discussion analytical and clear
- Conclusions rigorous and appropriate
- Suggestions for further study/identifies areas of uncertainty
- Abstract of research clearly presented at the start of the dissertation

Marker guidelines: Level of overall conclusions drawn from the body of work and grounded in theory and literature and student's own investigations. Evaluation should be done in a rigorous and appropriate manner. Meaningful recommendations for further study are made.

Presentation and clarity of expression

- Approach to the topic, ordering and coherence of sections
- Expression of thoughts and ideas
- Grammar and spelling
- Fluent academic writing style
- Appropriate citation of all references used (Harvard) in reference list

Marker guidelines: Appropriate presentation. Thoughts and ideas clearly expressed. Grammar and spelling accurate. A 2:1 or higher dissertation must have accurate and appropriately presented reference list.

Final dissertation: example marking scheme 2 (grid). The example marking grid is given in Table 9.6

Final dissertation: example marking scheme 3 (statement of achievement – criteria)

Students will achieve an excellent pass (1st class) if they can demonstrate:

- excellent range of data collection and thorough rigorous analysis, clearly identifying themes and patterns, illustrated appropriately;

Table 9.6 Example marking grid for dissertation

Criteria	1st class (70%+)	2.1 (60–69%)	2.2 (50–59%)	3rd class (40–49%)	Refer (0–39%)
Data analysis	Wide range of data collection. Excellent analysis of data, clearly identifying patterns and themes within the data collected. Critical analysis of data for reader. Where statistical analysis provided, wide range of appropriate graphs and charts to illustrate data. Where interviews and qualitative methods used, thorough critical analysis of respondent's views, creatively illustrated.	Reasonable range of data collection, good analysis of data, identifying patterns and themes within the data collected. Critical analysis of data for reader. Where statistical analysis provided, range of appropriate graphs and charts to illustrate data. Where interviews and qualitative methods used, critical analysis of respondents' views, well illustrated.	Limited range of data collected. Some analysis of data, some patterns and themes identified within the data collected. Some critical analysis of data for reader. Where statistical analysis provided, range of appropriate graphs and charts to illustrate data. Where interviews and qualitative methods used, some critical analysis of respondents' views with some illustration.	Very limited range of data collected. Limited analysis of data, very limited identification of patterns and themes within the data collected. Very limited critical analysis of data for reader. Where statistical analysis provided, minimal range of graphs and charts to illustrate data. Where interviews and qualitative methods used, limited critical analysis of respondents' views, more 'reportage' and storytelling, with few illustrations.	Minimal data collected. Little or no clear identification of patterns and themes within data. No critical analysis of data for reader. Where statistical analysis provided, inappropriate graphs and charts to illustrate data. Where interviews and qualitative methods used, no critical analysis of respondents' views, limited to reportage and mainly storytelling, little or no illustrations.
Depth of knowledge	Wide range of references (25+) of key texts (not using significant amounts of secondary referencing, unrefereed journal articles, Web data).	Reasonable range of references (20 max.) of key texts (using limited amounts of secondary referencing, unrefereed journal articles, Web data).	Limited range of references (10 to 15) of texts (using more secondary referencing, unrefereed journal articles, Web data).	Very limited range of references (<10) of texts (using significant amounts of secondary referencing, unrefereed journal articles, Web data).	References (<10) (using mainly secondary referencing, unrefereed journal articles, Web data).

Table 9.6 (Continued)

Criteria	1st class (70%+)	2.1 (60–69%)	2.2 (50–59%)	3rd class (40–49%)	Refer (0–39%)
	Thorough critical analysis of subject area for reader. Demonstration of thorough understanding of main issues in topic area.	Critical analysis of subject area for reader. Demonstration of understanding of main issues in topic area.	Reasonable attempt at critical analysis of subject area. Demonstration of some understanding of main issues in topic area.	Limited critical analysis of subject area. Demonstration of little understanding of issues in topic area.	No critical analysis of subject area. Demonstration of minimal understanding of issues in topic area.
Overall conclusions	Appropriate chapter/s clearly expressing the overall conclusions that may be drawn from the body of work. Clear conclusions, well grounded in theory and literature and in the student's own investigations. Evaluation carried out in a rigorous and appropriate manner. Meaningful recommendations for further study.	Appropriate chapter/s expressing the overall conclusions which may be drawn from the body of work. Good development shown in summary of arguments based in theory/literature and student's own investigations. Good evidence of evaluation carried out in dissertation. Appropriate recommendations for further study.	Chapter/s clearly expressing the overall conclusions. Evaluation reasonably well carried out. Evidence of findings and conclusions grounded in theory and literature and student's own investigations. Mainly appropriate recommendations for further study.	Chapter/s expressing some conclusions which may be drawn from the body of work. Some evaluation carried out. Limited evidence of findings and conclusions supported in theory and literature and student's own investigations. Few meaningful recommendations for further study.	Chapter/s expressing inappropriate, invalid, or unsubstantiated conclusions. No or incorrect evaluation carried out. No real investigations undertaken by student. No recommendations for further study.
Presentation	Presentation shows a polished and imaginative approach to the topic. Thoughts and ideas clearly expressed. Grammar and spelling accurate. Fluent academic writing style.	Presentation carefully and logically organised. Thoughts and ideas clearly expressed. Grammar and spelling accurate and language fluent.	Presentation satisfactory showing organisation and coherence. Language mainly fluent; grammar and spelling mainly accurate.	Presentation shows an attempt to organise in a logical manner. Meaning apparent but language not always fluent. Grammar and spelling contain errors.	Presentation is disorganised and incoherent. Purpose and meaning of dissertation and/or language is unclear. Grammar and spelling contain errors.

- use of excellent range of references with thorough critical analysis, demonstrating key issues in topic area;
- appropriate overall conclusions that are clear, well grounded in theory, literature and student's own investigations, with meaningful recommendations for further study;
- first class degrees require a high standard of referencing and marks will be deducted for failure to meet this requirement.

Students will achieve a good pass (2.1 or 2.2 degree classification) if they demonstrate:

- good range of data collection and good analysis, identifying themes and patterns, illustrated appropriately;
- use of good range of references with critical analysis demonstrating key issues in topic area;
- experienced presentation, clear expression, grammar and spelling.

Students will achieve a pass (3rd class degree level) if they can demonstrate:

- limited range of data collection and limited analysis, limited identification of themes and patterns, more reportage, limited illustration;
- limited range of references with limited critical analysis demonstrating little understanding of issues in topic area;
- some conclusions noted, limited evidence of conclusions being grounded in theory, literature and student's own investigations with few meaningful recommendations for further study;
- attempt to structure presentation, unclear expression, mistakes in grammar and spelling.

Students will be referred (achieve a mark below 40) if they demonstrate:

- minimal range of data collection and minimal analysis, no clear identification of themes and patterns, little or no illustration;
- very limited range of references with little or no critical analysis demonstrating minimal understanding of key issues in topic area;
- chapters expressing unsubstantiated or invalid conclusions, no grounding in theory or literature and no real student investigation. No recommendations for further study.
- disorganised and incoherent presentation, unclear expression, grammar and spelling errors.

Final dissertation: example marking scheme 4 (statement of achievement – strengths and weaknesses)

3rd class

- Written work of poor academic quality while maintaining meaning

- Referencing present but to lower than anticipated standard – for example, lacking in accuracy or missing citations
- Limited literature review

Overall

- A limited body of data collection
- Data presentation to poor standard
- Themes and patterns not fully identified
- No statistical analysis of primary data and/or
- Little or no analysis of interviews (e.g. using quotes only) and/or
- Discussion limited; fails to address all strands of research

2:2

- Should demonstrate a good quality of written work
- Referencing throughout text, but may be lacking in accuracy or missing citations

Overall

- A reasonable body of data collection
- Clear data presentation
- Themes and patterns identified

However

- Lacking detailed statistical analysis of primary data and/or
- Limited analysis of interviews (e.g. using quotes only) and/or
- Limited literature review (where this is the sole method) and analysis
- Discussion limited in scope or fails to address all strands of research together

2:1

- Should demonstrate a good quality of written work
- Should have clear referencing throughout text

Will demonstrate most or all of the following:

- A good body of data collection
- Clear data presentation
- Themes and patterns identified
- Statistical analysis of primary data and/or
- Detailed analysis of interviews using a coding methodology and/or
- Significant literature review (where this is the sole method) and analysis
- Contains a detailed discussion which draws all strands of research together

1st class

- Should demonstrate outstanding work
- Must have clear referencing throughout text

Will demonstrate exemplary levels of achievement in most or all of the following:

- Data collection and presentation
- Analysis of themes and patterns to very high standard
- High level of statistical analysis of primary data and/or
- Detailed analysis of interviews using a coding methodology and/or
- Significant literature review (where this is the sole method) and analysis
- Contains a detailed discussion that draws all strands of research together and contains insights which exceed those expected in a 2:1 dissertation

Final dissertation: example marking scheme 5 (summary)

Abstract
Identifies topic areas, issues investigated and the aims of the work. The information used and the form of the analysis is clearly stated. The main findings are presented and conclusions stated.

Literature review
An overview of material is presented in a structured format of appropriate breadth and depth.

Statement of the problem
A brief account of the underlying problem investigated is presented. The aim of the research is given.

Methodology
The research question or hypothesis is stated. Theoretical perspectives of the research are stated. The methods selected are identified and justified.

Results and discussion
An argument is presented which is grounded in literature and student findings. Critical evaluation is undertaken and weaknesses in the argument identified.

Conclusions and recommendations
The conclusion restates the aim and research question. The evidence is assessed and the research question answered. Recommendations for future work are made.

Presentation
Harvard system for referencing used accurately. Reference list is in alphabetical order without numbers or bullet points. The document is written in an appropriate style with attention to spelling and grammar.

Viva: example marking scheme 1 (list of criteria)

Depth and breadth of knowledge of topic area

- Depth and breadth of knowledge demonstrated
- Presentation interesting and stimulating

- Subject presented in clear and logical manner
- Subject presented effectively

Understanding of research methodology

- Demonstrated understanding of research theory
- Methods available clearly articulated
- Strengths and weaknesses highlighted as appropriate
- Clear linkages between methodology and proposal

Practical implications of conducting research

- Understood practical implications
- Selection of method clearly articulated and justified
- Logistical issues addressed
- Ethical issues addressed
- Reflection on progress

Ability to respond to probing questions

- Answers prompt and to the point
- Answers addressed the question being asked
- Responses defended dissertation choices
- Answers had appropriate level of detail

Supervisor guidance regarding writing up and assessment

Guidance from Supervisor A

'Writing up' should not be considered to be something that begins only when all other aspects of the research dissertation have been completed. It certainly should not be planned or viewed as a stand-alone activity, only to be started when all other aspects of the research project have run their course.

My view is that writing up is a process that should begin very early. It is a process that involves a number of constituent components, not just simply typing up what has been done and how it has been done along with an analysis, discussion and conclusion regarding findings.

I advise students to view writing up as an ongoing process which constitutes the following components: determining early in the dissertation process the possible architecture (or structure) of the dissertation thesis, documenting the aims, objectives and/or hypothesis, and gathering and filing literature in a structured manner relevant to the research objectives. These processes all support and contribute to writing, as does drafting, refining, then further developing and further refining a review of key literature pertinent to your chosen topic and research objectives. It is imperative that you share and discuss with your supervisor your early written reviews of literature along with your proposed research methodology.

It is very useful, when you are considering presentation, to take time to look at other people's completed dissertations. These should be made available to you by your university or college. The topics of the dissertations that you look at do not need to match your own research topic as it is the structure and presentation that you will be reviewing.

It is important to write, refine and logically file your writings as you progress through the dissertation journey. Taking such an approach to writing up will enable you, in the time approaching the submission deadline, to focus on editing, refining and enhancing the quality of your written thesis. This approach of 'writing as you go' can help to ensure that the quality of your written thesis does not suffer as a result of being rushed at the last minute. Set time aside regularly for writing up. Then make sure that your work is logically filed and backed up regularly so that the loss, corruption or failure of a memory stick or hard drive does not equate to the loss of all your hard work.

Ensure that you take time to analyse the assessment criteria by which your dissertation will be graded. The assessment criteria will detail expectations regarding presentation. If in doubt, make sure to ask your supervisor if she or he can show you an example of a well-presented dissertation. I show my students a 'well-presented' dissertation and, in brief discussion, relate the example to the presentation aspect of the assessment criteria. Dissertation presentation is invariably graded on aspects including: clear and appropriate structure of the submission, appropriate font and font sizing, clarity of expression, the inclusion of appropriate diagrams and illustrations, the correct sizing and labelling of diagrams and illustrations, the appropriate use of headers, footers and page numbering, and the suitable deployment of the correct system of referencing.

Appendices and references must not be included as an afterthought. Rather, appendices and sources listed in the reference section should only be included in your final document when they have been clearly referred to in the main text of the dissertation thesis.

Please do ensure that you actually read through your own work, and take time to read it out loud. It is very easy to miss one's own grammatical errors. Once you have read and refined your work then ensure that you give it to someone else to read through. This is asking quite a bit of somebody in terms of both their time and the responsibility placed upon them. This person should not be your supervisor; it is not your supervisor's job to proofread your work. Make sure to ask around and identify a potential proofreader well before you require their help. In this way, they will be prepared well in advance of the task and you will have set yourself an interim deadline for completion of your draft dissertation. Do remember to thank the person who proofreads your work and take on board their corrective feedback in the spirit that it is offered.

Finally, complete the write up and printing of your dissertation in good time before the submission deadline so as to allow for it to be suitably bound. Months before submission, you should investigate the possible options and requirements regarding binding. Please note that at busy times of the year, turnaround time for dissertation binding can vary and may greatly increase due to hundreds of students seeking binding at the same time. As such, do not simply expect to have your dissertation bound and returned to you within a day; this may not be feasible.

Guidance from Supervisor B

The key to a successful write up, in my view, is to start early and do a little bit of your write up on a regular basis. The first things I check are the references, working systematically from cover to cover. Clear use of good English is important. I hate reading documents where the student has used long and obscure words just to be impressive. Good graphs are really important. For me, pie charts and badly executed 3D graphs are ones I usually find in weaker pieces of work. I like the use of tables to summarise information or a legible visualisation, but not too many! Page numbering and tables of contents all need to tally. If you use MS Word then find out about how to use Styles such as Headings and how to insert captions to automate this.

Use storage on the cloud to allow you to work on your documents wherever you happen to be without risking losing your USB stick. Use clear version control, adopting a format for file names that identifies the chapter and the date of last editing. Save as a new name after each significant edit. Another advantage of working with split chapters is that you can send versions to your proofreader as you work on them for comments and feedback rather than handing a very large document to them at once. Your dissertation should be written in a format which allows the content to be understood by an educated but non-expert reader. If a peer from your course can't understand it then it is not explained adequately enough. Rewrite it.

If I was to provide a checklist for write-up, it would go something like this:

The work should be thoughtfully written, well structured with a good standard of presentation. The document should demonstrate technical competence and be clearly and logically structured.

The aims and objectives should be clearly stated.

The literature review should have an appropriate number of sources, clearly referenced and with the outcomes clearly stated. Mapping of key factors raised in papers may provide useful evidence and be a way of summarising a large body of information. Students should avoid having too few academic sources in the literature review; they should also avoid overuse of industry reports and trade magazines. No obvious sources of literature should be missing.

Methodology should be justified. There should be brief reflection on the methodology chosen, and the methods used in the questionnaire/interview/case study, etc. should link back to the literature.

Data should be obtained from appropriate sources. Ethical and/or health and safety considerations should be demonstrated.

There should be a clear interpretation and presentation of findings. Where statistical tests are used, these should be appropriate and carried out correctly with a statement of the findings from each test.

In general, too many small sections can undermine the flow of the work, as can too many graphs with no clearly stated purpose or multiple screenshots that may have been included for padding.

In the discussion section, a grid of the points and evidence from your work and how this links to the literature could be useful. The conclusion should link to the research question and should clearly address the achievement of the aims and objectives. Appropriate and relevant conclusions should be made, having been arrived at from clear and logical argument.

The conclusion should not be one that would have been obvious without doing any of the research work. The conclusion should clearly derive from the analysis of the data presented in the dissertation.

If using interviews, consider inserting one interview transcript into an appendix, showing how it was coded. More may be appropriate. These should be anonymised by default.

Recommendations for future work should be given.

Finally, the title should reflect the content of the work.

I often get asked about appendices when students begin writing up. Some things really have to go in the appendices, such as a summary of Excel sheets of lab data, your health and safety assessments and your ethical review. Bear in mind that a marker may not even look at your appendices and that even if they do, you may not receive academic credit. If you still feel that you want to leave it in the document because it's important to you then put it in the appendices.

Guidance from Supervisor C

When writing up a dissertation, you might like to think of a courtroom. In this, you will be the judge. Regardless of your thoughts and feelings about whether the defendant (your topic) in the dock is 'innocent' or 'guilty', the defendant deserves a fair trial. If you have any preconceived ideas, you should try to shelve them and consider only the evidence.

You may consider each of your methods or aspects of particular methods (e.g. the questionnaires, laboratory tests, etc.) as the witnesses. Call each one to the stand and allow the council for the prosecution and council for the defence to each have their say. As judge, you should summarise the key messages and present this for your jury (your reader) in the light of your understanding of the subject. Finally, your task is to sum up so that you can draw conclusions in such a way that your jury can see how these have been derived.

Make sure that your spelling and grammar is up to snuff. Get your document proofread by someone who knows what they are doing and is not afraid to cover your work with red pen. Work with someone from your course to review each other's work, or ask an independent person. Proofreaders must always be named in the acknowledgements. Work on the write-up systematically with your efforts increasing towards the deadline date so that the full scope of the project is clear in your mind. You should have a detailed understanding of the nature of the findings and the literature in your mind as you begin to write the discussion section. Give the pertinent facts and keep it concise. You should work towards the minimum word count rather than the maximum. A big bugbear of mine is the use of appendices either for padding or as a dumping ground for everything that was collected. What is the point in putting in printouts of everything, copies of questionnaires, copies of web pages? It's an absolute waste of time and paper. If it is not important enough to be discussed in the main document then it should not be in the document.

Student views on writing up and assessment

'START EARLY – don't leave it until the last minute!!! Write a bit each week to stay on top of things. Make sure you write during any holiday breaks ... it will pay off in the end!'

'Get the structure defined first; then run this by your supervisor for comments. Start to flesh out the write up over a period of a few months, building on your proposal document. Ensure that the dissertation flows. When it's finished, proofread it several times, and on the last proofread, take about three to four hours; this will allow for any reduction/increase in word count necessary and checking of spelling/grammar, etc.'

'Stay on top of writing; don't leave it and get bogged down.'

'Definitely choose something you have an interest in as a lot of time is spent on researching and writing about that topic.'

'Make sure that you constantly work at researching and writing. Do not leave it for longer than a week or you will lose track of where you are. I continually worked on it and constantly read over the progress. As a result,

I managed to hand it in three days early and was far less stressed than many of my peers.'

'Keep on top of references from the start. Make sure you do your reference list at the same time as your literature review – this ensured I didn't have to do my list at the last minute. Bear in mind how long the document will take to format/put together. You should definitely allow a whole day just for formatting and tidying up. Seriously, fully reference your sources as you use them. This will save you hours when deadline is looming. (My housemate was up until 7 a.m. the night beforehand doing this!)'

'I don't think most people start their final year thinking things are going to go wrong. I had a family bereavement in the January and that completely threw everything into chaos. I think if I hadn't been aware of the correct procedures, I would have come completely unstuck. Some universities have procedures in place to help you; others just say "it's part of life – tough". Luckily I was registered at the former! As it was, getting restarted on the work, when I was able to, was a big uphill struggle.'

'I started to write up mine by laying out the sections I needed; for example, Introduction, Methodology, Literature Review, Case Studies, etc. This helped me as I could fill out each section and then see what I still had to finish. I ensured all my data was collected before the final write up; that way, I only put in relevant material and could structure the dissertation more effectively. Each time I completed a section, I read through it and changed things if I needed to. It really saved time in the long run.'

'Don't complain about having to do the work. Everyone has to do it so just get on with it; before you know it, you will be looking for something to delete!'

'Do not listen to what other people have done in their dissertation as it will only have you doubting your own. It is your research work so the way you want to do it is right.'

'Get into the hang of using formats in MS Word. Again, the features such as producing contents, lists of figures and tables, etc. will save you a lot of time and stress.'

'I would advise you to create a timetable and stick to it; this way, via planning ahead, you will be fine.'

'Use an action plan to plan the work and hit targets. Aim to have the primary data collection done in time to allow at least three months of analysis and write up. Writing up will take longer than you think. Word count doesn't matter until the last few drafts; aim high and reduce words rather than the other way around.'

'Can be thoroughly enjoyable if you choose a topic of interest; make the effort to fully read around the subject – do not make your results up. You will find it hard to get motivated about the findings and your work will suffer if this is the case. I'm obviously not speaking from personal experience here!'

'Personally I have really enjoyed the dissertation; it is a lot of work but, on reflection, it isn't something to dread.'

'It was a very daunting experience, especially at the start when I didn't fully understand the expectations of a dissertation. But as time passed, everything became clearer. I actually enjoyed doing the work and I am glad I opted to do dissertation. A very satisfying experience once the work is done!'

Summary

This chapter has drawn attention to important issues concerning the writing up and the assessment of the dissertation. The chapter has considered the structure and contents of a final dissertation thesis and has drawn attention to the need to think about writing style when writing up. Issues of plagiarism and academic misconduct have also been addressed.

Assessment of the dissertation has been another key consideration of this chapter. The *viva voce*, the oral assessment of the dissertation, has been discussed with particular regard to when it might be carried out, how the researcher might prepare for it and typically what questions might be asked. The assessments of the dissertation proposal and the final dissertation thesis have also been considered, with various examples presented. Finally, the chapter provided suggestions for further reading, guidance from three supervisors and some brief student reflections regarding writing up and assessment.

By considering the issues that have been raised in both this chapter and the book and by following the guidance provided, research students will be better supported throughout the dissertation journey.

Once writing up, proofreading, document binding, submission and assessment have occurred, it only remains for the dissertation researcher to plan their celebration and reflect upon what they might do differently next time!

Suggested further reading

ALLISON, B. & RACE, P. (2004) *The Student's Guide to Preparing Dissertations and Theses*. 2nd ed. London: Routledge Falmer. ISBN 0-415-33486-1

BELL, J. (2010) *Doing Your Research Project: A guide for first-time researchers in education, health and social science*. 5th ed. Maidenhead: McGraw-Hill Open University Press. ISBN 978-033523582-7

CRESWELL, J. W. (2012) *Educational Research: Planning, conducting, and evaluating quantitative and qualitative research*. 4th ed. Boston MA: Pearson. ISBN 978-0-13-261394-1

DAWSON, C. (2009) *Introduction to Research Methods: A practical guide for anyone undertaking a research project*. 4th ed. Glasgow: Bell & Bain Ltd. ISBN 978-1-84528-367-4

FELLOWS, R. & LIU, A. (2008) *Research Methods for Construction*. 3rd ed. Chichester: Wiley-Blackwell. ISBN 978-1-4051-7790-0

FISHER, C. et al. (2010) *Researching and Writing a Dissertation: A guidebook for business students*. 3rd ed. Harlow, Essex: Pearson Education Limited. ISBN 978-0-273-72343-1

GLATTHORN, A. A. & JOYNER, R. L. (2005) *Writing the Winning Thesis or Dissertation: A step-by-step guide.* Thousand Oaks, CA: Corwin Press. ISBN 0-7619-3961-X

GREETHAM, B. (2009) *How to Write Your Undergraduate Dissertation.* Houndmills: Palgrave Macmillan. ISBN 978-0-230-21875-8

HOLT, G. D. (1998) *Guide to Successful Dissertation Study for Students of the Built Environment.* 2nd ed. Wolverhampton: Built Environment Research Unit, University of Wolverhampton. ISBN 1-902010-01-9

NAOUM, S. G. (2007) *Dissertation Research and Writing for Construction Students.* 2nd ed. Oxford: Butterworth-Heinemann. ISBN 0-7506-8264-7

SWETNAM, D. (2001) *Writing Your Dissertation: How to plan, prepare and present successful work.* 3rd ed. Oxford: How To Books Ltd. ISBN 1-85703-662-X

WALLIMAN, N. (2011) *Research Methods: The basics.* Abingdon: Routledge. ISBN 978-0-415-48994-2

Appendix: example calculations

Chi square calculation using MS Excel

	A	B	C	D	E	F	G
1		Observed Frequencies (Collected data)					
2		Category 1	Category 2	sum			
3	Group A	56	72	128	=SUM(B3:C3)		
4	Group B	14	19	33	=SUM(B4:C4)		
5	Group C	36	34	70	=SUM(B5:C5)		
6	sum	106	125	231	=SUM(B3:C5)		
7		=SUM(B3:B5)	=SUM(C3:C5)				
8		Expected Frequencies (Calculated)					
9		Category 1	Category 2				
10	Group A	58.74	69.26		=B6/$D$6*D3/$D$6*$D$6	=C6/$D$6*D3/$D$6*$D$6	
11	Group B	15.14	17.86		=B6/$D$6*D4/$D$6*$D$6	=C6/$D$6*D4/$D$6*$D$6	
12	Group C	32.12	37.88		=B6/$D$6*D5/$D$6*$D$6	=C6/$D$6*D5/$D$6*$D$6	
13	(note totals of columns and rows should be the same as Observed data)						
14	Calculation of x^2						
15	O	E	$(O-E)^2/E$		$(O-E)^2/E$		
16	56	58.74	0.13		=((A17-B17)^2)/B17		
17	14	15.14	0.09		=((A18-B18)^2)/B18		
18	36	32.12	0.47		=((A19-B19)^2)/B19		
19	72	69.26	0.11		=((A20-B20)^2)/B20		
20	19	17.86	0.07		=((A21-B21)^2)/B21		
21	34	37.88	0.40		=((A22-B22)^2)/B22		
22			x^2	1.26			
23	degrees of freedom (DF			2	=(2-1)*(3-1)		
24					= number of rows -1 X no of columns - 1		
25	level of significance			5%			
26							

Figure A.1 Screenshot showing calculations for chi square

Procedure

Refer to Figure A.1.

1. Type in the data under Category 1 and 2, Group A to C.
2. Calculate the sum of the rows in E2, E4 and E5.
3. Calculate the sum of columns in B7 and B8.

Critical values of χ^2

DF	0.05			
1	3.84			
2	5.99			
3	7.82			
4	9.49			
5	11.07			
6	12.59			
7	14.07			
8	15.51			
9	16.92			
10	18.31			
11	19.68	DF		2
12	21.03	table value		5.99
13	22.36	calculated χ^2		1.26
14	23.69	**Accept H0, not significant**		
15	25.00			
16	26.30			
17	27.59			
18	28.87			
19	30.14			
20	31.41			

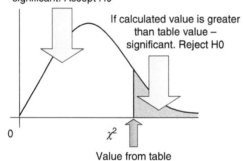

If calculated value is less than table value – not significant. Accept H0

If calculated value is greater than table value – significant. Reject H0

0

χ^2

Value from table

Figure A.2 Using chi square table values to establish outcome

4. Calculate the expected frequencies in B10, B11, C10, C11, D10, D11. The equations are shown in columns E and F. The sums of these rows and columns should be the same as the original data.
5. In cell A16, type =B3 and copy down.
6. In cell B16, type =B10 and copy down.
7. Repeat for the remaining data so that the observed and expected cells are being compared with each other.
8. Type the formula shown in E16 into C16.

9. The sum of this column is the chi square statistic.
10. Work out the degrees of freedom:

$$df = (\text{number of rows} - 1) \times (\text{number of columns} - 1)$$
$$= (2 - 1) \times (3 - 1)$$
$$= 2$$

11. Decide the level of significance (e.g. 5 per cent).
12. Using *df*, find the critical value from the table (refer to Figure A.2).
13. State conclusion.

If the number calculated for x^2 is greater than the critical x^2 in the table, H0 is rejected and H1, that there is a significant difference in distribution, is accepted.

If the number calculated for x^2 is smaller than the critical x^2 in the table, H0, that there is no significant difference in distribution, is accepted.

Chi square with Yates' correction using MS Excel

	A	B	C	D	E	F	G
1		Observed Frequencies (Collected data)					
2		Category 1	Category 2	sum			
3	Group A	189	67	256	=SUM(B3:C3)		
4	Group B	25	26	51	=SUM(B4:C4)		
5	sum	214	93	307	=SUM(B3:C4)		
6		=SUM(B3:B4)	=SUM(C3:C4)				
7		Expected Frequencies (Calculated)					
8		Category 1	Category 2				
9	Group A	178.45	77.55		=B5/$D$5*D3/$D$5*$D$5	=C5/$D$5*D3/$D$5*$D$5	
10	Group B	35.55	15.45		=B5/$D$5*D4/$D$5*$D$5	=C5/$D$5*D4/$D$5*$D$5	
11	(note totals of columns and rows should be the same as Observed data)						
12		Calculation of x^2				Formula for	Formula for
13	O	E	(O-E)	Yates	$(O-E)^2/E$	Yates correction	$(O-E)^2/E$
14	189	178.45	10.55	10.0505		0.566 =IF(C14<0,C14+0.5,C14-0.5)	=(D14^2)/B14
15	25	35.55	-10.55	-10.0505		2.841 =IF(C15<0,C15+0.5,C15-0.5)	=(D15^2)/B15
16	67	77.55	-10.55	-10.0505		1.303 =IF(C16<0,C16+0.5,C16-0.5)	=(D16^2)/B16
17	26	15.45	10.55	10.0505		6.538 =IF(C17<0,C17+0.5,C17-0.5)	=(D17^2)/B17
18					x^2	11.25	
19							
20	degrees of freedom (D		1		=(2-1)*(2-1)		
21					= number of rows -1 X no of columns - 1		
22	level of significance		5%				
23							
24							

Figure A.3 Screenshot showing calculations for chi square with Yates' correction for two by two grid

Procedure

Refer to Figure A.3.

1. Type in the data under Category 1 and 2, Group A and B.
2. Calculate the sum of the rows in E2 and E4.
3. Calculate the sum of columns in B7 and B8.
4. Calculate the expected frequencies in B9, B10, C9, C10. The equations are

shown in columns E and F. The sums of these rows and columns should be the same as the original data.

5. In cell A14, type =B3 and copy down.
6. In cell B14, type =B9 and copy down.
7. Repeat for the remaining data so that the observed and expected cells are being compared with each other.
8. Calculate O–E for each row starting with =A14-B14 in cell C14.
9. Type the formula shown for Yates' correction in F14 into D14.
10. Type the formula shown in G14 into E14.
11. The sum of column E is the chi square statistic.
12. Work out the degrees of freedom:

df = (number of rows – 1) × (no of columns – 1)
   = (2 – 1) × (2 –1)
   = 1

13. Decide the level of significance (e.g. 5 per cent).
14. Using df, find the critical value from the table (refer to Figure A.4).
15. State conclusion.

If the number calculated for x^2 is greater than the critical x^2 in the table, H0 is rejected and H1, that there is a significant difference in distribution, is accepted.

If the number calculated for x^2 is smaller than the critical x^2 in the table, H0, that there is no significant difference in distribution, is accepted.

ANOVA (one-way independent measures) using MS Excel ToolPack

Single factor ANOVA (analysis of variance) is used to test the null hypothesis that the means of several populations are all equal.

HO: all means are the same
H1: at least one of the means is different

Procedure

1. In MS Excel, install the Analysis ToolPack.
2. Click the Microsoft Office button.
3. Select Add-ins to open a box.
4. Select Analysis ToolPack and click OK.
5. Type in the data shown in Figure A.5.
6. On the Data tab, click Data Analysis.
7. Select ANOVA: Single Factor and click OK.
8. Enter the Input Range box and select the range A1:C1.
9. Select Grouped by Columns.
10. Tick Labels in first row.
11. Alpha 0.05.

Critical values of χ^2

DF	0.05		
1	3.84		
2	5.99		
3	7.82		
4	9.49		
5	11.07		
6	12.59		
7	14.07		
8	15.51		
9	16.92		
10	18.31		
11	19.68	DF	2
12	21.03	table value	3.84
13	22.36	calculated χ^2	11.25
14	23.69	**Accept H1, significant**	
15	25.00		
16	26.30		
17	27.59		
18	28.87		
19	30.14		
20	31.41		

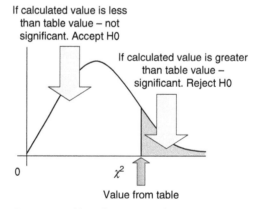

Figure A.4 Using chi square table values to establish outcome

12. Output options: Select output range and enter E1 and click OK.
13. See Figure A.6 and state the conclusion of the test.

If the number calculated for F is greater than the critical F, the null hypothesis is rejected and the conclusion is that at least one of the means is different.

As calculated, F (10.4075) is greater than F-crit (3.354) at 5 per cent level of significance. H0 is rejected and H1 accepted, and it is concluded that at least one of the means is different.

	A	B	C	D
1	Set A	Set B	Set C	
2	5	5	10	
3	1	5	7	
4	5	7	9	
5	4	5	6	
6	3	8	7	
7	1	4	10	
8	5	5	9	
9	4	3	4	
10	1	4	3	
11	5		10	
12			6	
13				
14				

Figure A.5 Three data sets

	E	F	G	H	I	J	K
1	Anova: Single Factor						
2							
3	SUMMARY						
4	Groups	Count	Sum	Average	Variance		
5	Set A	10	34	3.4	3.155556		
6	Set B	9	46	5.111111	2.361111		
7	Set C	11	81	7.363636	6.054545		
8							
9							
10							
11	Source of Variation	SS	df	MS	F	P-value	F crit
12	Between Groups	83.13232323	2	41.56616	10.4075	0.000446	3.354131
13	Within Groups	107.8343434	27	3.993865			
14							
15	Total	190.9666667	29				
16							
17							

Figure A.6 ANOVA single factor screenshots

Note: ANOVA does not identify which of the means is different. For each of the sets, *t*-tests will need to be done as a post-test.

Binomial test using MS Excel

A student found literature which stated that 80 per cent of homes had smoke detection. The amount of detection was found to be higher year on year. The questionnaire used found that 95 per cent of homes had smoke detection.

H0: amount of detection no different to that in the literature

H1: amount of detection higher than that suggested in the literature (one-tailed)

Procedure

Refer to Figure A.7

1. Enter labels into A3 and A4 and data into B3 and B4.
2. Calculate sum in cell B5.

	A	B	C	D	E	F	G
1	Is smoke detection present?						
2							
3	yes	71					
4	no	4					
5		75	total respondants				
6							
7		0.8	probability given in literature				
8		60	np	=B7*B5			
9			responses which would be anticipated				
10							
11		3.464	estimate of standard deviation				
12			$=(np(1-p))^{0.5}$				
13			=(B5*B7*(1-B7))^0.5				
14			is 71 higher than would be expected				
15		70.5	x	=B3-0.5			
16							
17		3.031089	z	=(B14-B8)/B11			
18		0.998782	Φ (z)	=NORMSDIST(B17)			
19							
20							
21							
22							
23							
24							
25							
26							
27							
28							
29							
30							

Figure A.7 Three data sets

3. Enter the value 0.8 into B7.
4. Calculate np in B8 using the equation given in D8.
5. Estimate standard deviation in B11 using the equation given in D11.
6. Calculate the critical point x in B14.
7. Calculate z.
8. Calculate $\phi(z)$ – the probability.
9. Compare with critical values:

$z_{0.025}$ = 1.96 (critical value two-tailed)
$z_{0.05}$ = 1.64 (critical value one-tailed)

H0 is rejected ($3.03 \geq 1.64$ at $\alpha = 0.05$) and it is concluded that there is a significant increase in the number of smoke detectors observed in the study.

Index